THE ELEMENTS

						VIIA	O
		IIIA	IVA	VA	VIA	1 **H** 1.00797	2 **He** 4.0026
		5 **B** 10.811	6 **C** 12.01115	7 **N** 14.0067	8 **O** 15.9994	9 **F** 18.9984	10 **Ne** 20.183
IB	IIB	13 **Al** 26.9815	14 **Si** 28.086	15 **P** 30.9738	16 **S** 32.064	17 **Cl** 35.453	18 **Ar** 39.948
28 **Ni** 58.71	29 **Cu** 63.54	30 **Zn** 65.37	31 **Ga** 69.72	32 **Ge** 72.59	33 **As** 74.9216	34 **Se** 78.96	35 **Br** 79.909	36 **Kr** 83.80
46 **Pd** 106.4	47 **Ag** 107.870	48 **Cd** 112.40	49 **In** 114.82	50 **Sn** 118.69	51 **Sb** 121.75	52 **Te** 127.60	53 **I** 126.9044	54 **Xe** 131.30
78 **Pt** 195.09	79 **Au** 196.967	80 **Hg** 200.59	81 **Tl** 204.37	82 **Pb** 207.19	83 **Bi** 208.980	84 **Po** (210)	85 **At** (210)	86 **Rn** (222)

63 **Eu** 151.96	64 **Gd** 157.25	65 **Tb** 158.924	66 **Dy** 162.50	67 **Ho** 164.930	68 **Er** 167.26	69 **Tm** 168.934	70 **Yb** 173.04	71 **Lu** 174.97
95 **Am** (243)	96 **Cm** (247)	97 **Bk** (247)	98 **Cf** (249)	99 **Es** (254)	100 **Fm** (253)	101 **Md** (256)	102 **No** (253)	103 **Lw** (257)

Atomic Weights are based on C^{12}—12.0000 and Conform to the 1961 Values

Second Edition

SYNTHESIS AND TECHNIQUE IN INORGANIC CHEMISTRY

Robert J. Angelici

Department of Chemistry
Iowa State University
Ames, Iowa

 **SAUNDERS GOLDEN SUNBURST SERIES**

1977
W. B. SAUNDERS COMPANY
Philadelphia, London, Toronto

W. B. Saunders Company: West Washington Square
Philadelphia, PA   19105

1 St. Anne's Road
Eastbourne, East Sussex BN21 3UN, England

833 Oxford Street
Toronto, Ontario M8Z 5T9, Canada

Library of Congress Cataloging in Publication Data

Angelici, Robert J

Synthesis and technique in inorganic chemistry.

(Saunders golden sunburst series)

Includes bibliographies and indexes.

1. Chemistry, Inorganic—Laboratory manuals.
   I. Title.   [DNLM: 1. Chemistry—Laboratory manuals.
   QD155 A582s]

QD155.A53 1977   546'.028   76-4244

ISBN 0-7216-1281-4

Synthesis and Technique in Inorganic Chemistry   ISBN   0-7216-1281-4

Last digit is the print number: 9   8   7   6   5   4   3   2   1

to my Mother and Father

Preface to the Second Edition

   As in the first edition, the intent of this revision is to instruct students in the modern synthetic and instrumental techniques currently used in inorganic chemistry. All of the experiments included in the first edition are retained in this one also. However, improvements have been made in many; in some cases this has also reduced the time required to complete the experiment. In Experiment 14, a new synthesis introducing the use of a high pressure autoclave (i.e., bomb) has been added. New questions have been added at the end of each experiment, and new references associated with each experiment have been included.

   At the end of each experiment has been added a new section called "Independent Studies." This section grew out of a desire to introduce the student to a more research-type investigation. The intent is to have the student extend the techniques learned in the experiment to other related systems. The idea is not to develop a long-term research program, but a short yet independent project.

   For example, Experiment 2 is concerned with determining the rate law for the aquation reaction,

$$Co(NH_3)_5Cl^{2+} + H_2O \xrightarrow{H^+} Co(NH_3)_5(OH_2)^{3+} + Cl^-, \text{ at } 60°C$$

As an independent study, the student might investigate the rate of the reaction at different temperatures and calculate the activation parameters, or study it at different ionic strengths, or measure the equilibrium constant for the reaction, or determine the catalytic effect of $Hg(NO_3)_2$ on the reaction, or study the rate of aquation of $Co(NH_3)_5Br^{2+}$ to compare it with that of $Co(NH_3)_5Cl^{2+}$. These suggestions listed in the "Independent Studies" section are intended to encourage the student to think about interesting extensions of the regular experiment.

   In selecting an independent study, the student must consult the literature to determine what chemicals and equipment are required, plan the experiment, and evaluate the results independently. In our experience it has been a very worthwhile addition to the course, challenging the student to undertake his or her own project. As in any research, many are successful in achieving their goal, while others are not. The number of independent studies that can be carried out during the course will depend upon the time available; in our one-quarter course we have generally taken time for one or two independent projects.

Another addition to this edition is the series of ten Appendices at the end of the book, which give frequently used data such as properties of common solvents, transmission ranges of infrared and UV-visible cell window materials, drying agents, stable isotopes and their nuclear spins, and heating and cooling baths.

Comments from those who used the first edition have been very helpful in writing this revision. I am especially grateful to Dr. B. Duane Dombek for improvements in the experiments and to Professor Keith Purcell for his detailed comments on this edition. The encouragement of my wife, Libby, and my children, Scott and Karen, is very much appreciated.

ROBERT J. ANGELICI

Preface to the First Edition ━━━━━━━━━

Interest and research in inorganic chemistry have expanded in recent years. This has been partially a result of the discovery of new and exotic classes of compounds such as the inert gases, boron hydrides, and metal clusters. Probably of more importance is the recognition that inorganic chemistry is fundamental to vast areas of organic, physical, and biochemistry. These developments have largely stemmed from the synthesis and the characterization in considerable detail of new compounds.

Realizing that our students, upon graduation, would be doing laboratory work with the most modern of facilities in either graduate school or industry, we designed a junior-senior level inorganic laboratory course to prepare them for their future encounter with research. To be of value, the course had to include techniques which are used in chemical research today. Most of the techniques which we think are important have been included in our course and in this manual:

a. Inorganic synthetic techniques—for example, those requiring a dry box or bag, a vacuum line, very high temperatures, nonaqueous solvents such as NH_3 and N_2O_4, and electrolytic oxidation.

b. Methods of purification—ion exchange, thin layer and column chromatography, vacuum sublimation, distillation, recrystallization, and extraction.

c. Methods of characterization—infrared, ultraviolet-visible and nuclear magnetic resonance spectroscopy, mass spectrometry, ionic conductivity, optical rotations, magnetic susceptibilities, chemical reactivity and rates of reaction, and equilibrium constants for complex formation.

The experiments have been designed to illustrate the fundamentals of these techniques. It is assumed that if refinements or slight modifications of these techniques are required in any future research, they can be made by consulting the specialized references on techniques given at the end of each experiment.

The selection of experiments has been governed by several practical considerations. First, inherently hazardous procedures have been avoided, although lack of understanding or ignorance of proper methods can make any of the experiments dangerous. Thus careful attention should be paid to the safety notes given in the instructions. Second, for those institutions, such as ours, which have scheduled 3-hour laboratory periods the experiments have been selected to allow experimental work in 3-hour segments. The experiments which require longer periods are noted in the text. Third, we have limited the preparations to those which require relatively inexpensive chemicals. Finally, we felt that it was highly desirable for the student to prepare compounds which are either not commercially available or are very expensive. This again is an attempt to place the student in the position of a research chemist who would obviously buy needed compounds if they could be purchased at reasonable prices from commercial sources. This means that the syntheses in this manual are largely of compounds which were only recently prepared and which border on the frontiers of research.

In pursuing the above goals, the experiments have necessarily involved preparations and characterizations of a variety of types of compounds. Hence syntheses of nonmetal, transition metal, and organometallic compounds are represented. The experiments are all written in considerable detail so that the student may learn new techniques properly and as rapidly as possible. Many of the results, such as infrared and nuclear magnetic resonance spectra, optical rotations, mass spectra, and magnetic susceptibilities, will require interpretation. This interpretation will lead the student to the library to examine original papers and standard reference works. This would again place the student in the shoes of a researcher, and indeed it would be very desirable, if time permits, to extend his research to carrying out a synthesis which he himself would select from the literature.

During the past four years at Iowa State, this one-quarter course has involved two three-hour laboratory periods each week. Since the course is offered at the junior-senior level, all students had completed physical, analytical, and organic chemistry laboratory courses and were taking an independent lecture course in inorganic chemistry concurrently. Essentially the same course was available for first-year graduate students. During the one-quarter term, it was possible to complete approximately eight of the experiments. Since facilities vary from one institution to another and very few institutions will have equipment to do all the experiments given in this manual, a sufficient range of experiments is provided to create a meaningful course at almost all colleges and universities. Because the lack of nuclear magnetic resonance and mass spectral facilities is a very common problem, these spectra have been included at the back of the book for compounds which are amenable to characterization by these techniques. These can then be interpreted by the student. If NMR and mass spectral equipment are available, it would of course be more valuable to have the student obtain a spectrum on his own preparation.

In general, there is no required order for performing the experiments. To make the most efficient use of our facilities it was necessary to have each student perform a different experiment during a given laboratory period. Hence, experiments from throughout the manual were being carried out even during the first week of the course. The only restriction on the ordering of experiments results from the dependence of some measurements upon a previous preparation, such as the kinetic study of the aquation of $Co(NH_3)_5Cl^{2+}$ to $Co(NH_3)_5(OH_2)^{3+}$ which requires the prior synthesis of $Co(NH_3)_5Cl^{2+}$. In this case and others, some of the necessary compounds may be purchased commercially if desired. Hints on the organization of the laboratory and on commercial sources for special equipment and chemicals are given in Notes to the Instructor.

Our experience with this laboratory course has been a particularly exciting and gratifying one. It is in large measure the enthusiasm of the students which has led to the writing of this book. For suggestions and considerable help in the development of the experiments, I am very much indebted to J. Graham, R. Bertrand, G. McEwen, Dr. J. Espenson, and especially D. Allison and D. White. For the typing of the experiments, I am grateful to Mrs. L. Gustafson, Mrs. P. Feikema, and Mrs. L. Dayton. My gratitude goes to Forrest Hentz at North Carolina State University and Robert Kiser at the University of Kentucky for their reviews and suggestions on the final manuscript. Finally, I want to thank Iowa State University for providing me with the opportunity to develop and organize the experiments in this book.

Robert J. Angelici

Contents

Introduction

This course is designed to introduce the student to modern research techniques in inorganic chemistry. The experiments to be carried out involve the synthesis of various types of compounds by diverse experimental techniques. Modern instrumental methods will be used to characterize the products. The entrance into these new areas of research requires a greater appreciation of safety hazards related not only to the chemical properties of reactants but also to the dangers presented by unfamiliar apparatus. New fundamental techniques will be encountered, and these must become as second nature as weighing and pipetting were in prior courses. Finally, the research orientation of the course requires careful record-keeping by the researcher. This first chapter will expand upon certain aspects of (1) safety, (2) basic laboratory procedures, and (3) methods of keeping a research notebook. It is important that you become well acquainted with this material before beginning the experiments.

SAFETY

Since the chemicals and equipment used in this course can be extremely dangerous if handled improperly, it is very important that you be completely familiar, with the experiment before beginning it. This includes knowing the toxicity and reactivity of the chemicals used and the hazards related to the apparatus. This knowledge permits you to determine what is and what is not a dangerous situation. *Before undertaking any hazardous operation, know beforehand what you would do in case of an accident.* In case of fire, you should be familiar with the location and operation of fire extinguishers; a fire blanket should be on hand to be wrapped around persons whose clothing catches fire; the location of safety showers and eyewashers should be known and their use demonstrated; and a well stocked first aid kit (see references for recommended kit contents) should be available.

It is, of course, best to avoid accidents altogether by following the experimental procedures meticulously and by understanding the dangerous operations in an experiment. But since some accidents will nevertheless happen, the next best step is to take basic precautions to prevent

1

them from inflicting serious injury. *Eye protection is absolutely necessary at all times;* safety goggles protect you not only from your own experiment but also from your neighbor's. A laboratory apron or nonflammable coat should always be worn, and while pouring corrosive liquids protect yourself against spills by wearing rubber or polyethylene gloves. An appropriate shield should be placed between you and a potentially explosive reaction mixture. (Such explosive situations have been carefully avoided in this course, but Dewar vacuum flasks may implode with serious consequences. If Dewars are wrapped on the outside with a sturdy tape, implosions are considerably less dangerous.) Compressed gas cylinders should always be securely anchored to a wall or heavy bench. If a large cylinder tips over and the valve snaps off, the cylinder becomes a jet-propelled rocket which has sufficient power to penetrate a brick laboratory wall.

Chemicals should be assumed to be toxic unless known to be otherwise. Solids are relatively harmless because they can only be taken into the body orally or through open cuts on the hands and arms. If you refrain from smoking or eating in the laboratory, use gloves when necessary, and wash your hands after each laboratory period, the toxicity of solids will not be a hazard. On the other hand, vapors of volatile liquids and gases are much more dangerous, and their presence is frequently difficult to detect.

In Table I are listed Threshold Limit Values (TLV) in ppm of the toxic gas in air at 25°C and 760 mm Hg pressure. The Threshold Limit Values are concentrations of gas to which nearly all workers could be exposed for periods of months without adverse effect. Somewhat higher concentrations could be experienced for shorter periods. Some chemicals are fast acting and short exposures may cause serious effects, including death. These are given Ceiling values, designated by C in Table I. In these cases, the student should not be exposed to concentrations (in ppm) greater than that indicated by C, even for short periods. Unfortunately, most research laboratories do not have facilities for measuring concentrations of toxic gases in the air. For this reason, the numbers in Table I primarily serve to indicate relative toxicities. For example, the exceedingly toxic hydrogen cyanide, HCN, has a TLV of 10, which suggests that other substances with values of 10 or less, such as AsH_3, BF_3, F_2, HCl, $Ni(CO)_4$, NO_2, and SO_2, will also be very toxic. Moreover, these substances are very volatile and likely to give high gas concentrations. On the other hand, chemicals such as ethyl alcohol are relatively harmless because of their large TLV (1000) and also their low volatility.

The numbers in Table I are meant to serve only as a rough guide. To be sure, it is experimentally very difficult to establish meaningful safe concentration levels, and the toxicology of relatively few of the known chemicals has been evaluated. Since chemical research involves the preparation of new compounds whose toxic effects are totally un-

TABLE I   **Threshold Limit Values (TLV) of Toxic Volatile Liquids and Gases***

	TLV† *(in ppm)*		*TLV†* *(in ppm)*
Acetic acid	10	Ethyl alcohol	1000
Acetic anhydride	5	Ethylamine	10
Acetone	1000	Ethylenediamine	10
Acetonitrile	40	Fluorine	0.1
Ammonia	50	Formaldehyde	C 5
Aniline	5	n-Hexane	500
Arsine	0.05	Hydrazine	1
Benzene (Benzol)	C 25	Hydrogen bromide anhydrous	3
Boron trifluoride	C 1	Hydrogen chloride	C 5
Bromine	0.1	Hydrogen cyanide	10
Bromoethane		Hydrogen fluoride	3
(ethyl bromide)	C 200	Hydrogen peroxide, 90%	1
Bromomethane		Hydrogen selenide	0.05
(methyl bromide)	C 20	Hydrogen sulfide	10
1,3 Butadiene	1000	Iodine	C 0.1
n-Butylamine	C 5	Methanethiol	
Carbon dioxide	5000	(Methyl mercaptan)	10
Carbon disulfide	20	Methanol (Methyl alcohol)	200
Carbon monoxide	50	Nickel carbonyl	0.001
Carbon tetrachloride	10	Nitric Acid	2
Chlorine	C 1	Nitric oxide	25
Chlorine trifluoride	C 0.1	Nitrobenzene	1
Chlorobenzene	75	Nitrogen dioxide	C 5
Chloroethane	1000	Nitrogen trifluoride	10
Chloroform	C 50	Nitromethane	100
Cyclohexane	300	Oxygen difluoride	0.05
Cyclohexene	300	Ozone	0.01
Cyclopentadiene	75	Pentaborane	0.005
Decaborane	0.05	Perchloryl fluoride	3
Diborane	0.1	Phenol	5
p-Dichlorobenzene	75	Phosgene (Carbonyl chloride)	0.1
1,2-Dichloroethane		Phosphine	0.3
(ethylene dichloride)	50	Phosphorus trichloride	0.5
Dichloromethane		Pyridine	5
(methylene chloride)	500	Selenium hexafluoride	0.05
Diethylamine	25	Sulfur dioxide	5
Diethyl ether	400	Sulfur hexafluoride	1000
N,N-Dimethylformamide	10	Sulfuryl fluoride	5
Dimethyl sulfate	1	Tellurium hexafluoride	0.02
p-Dioxane	100	Tetrahydrofuran	200
Ethanethiol		Toluene	200
(ethyl mercaptan)	C 10	Triethylamine	25
Ethyl acetate	400	p-Xylene	100

*N. V. Steere, ed., *J. Chem. Educ., 44,* A45 (1967).
†Ceiling short-duration exposure limits are designated by C.

known, it is best to establish safe laboratory practices based on the assumption that all chemicals are toxic, particularly those that vaporize readily. If toxic volatile liquids and gases are always handled carefully in an efficient hood, their toxicity should never produce a dangerous

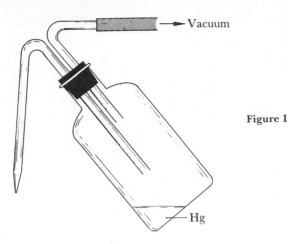

Figure I

situation. *When in doubt about the toxicity of reactants or products in a reaction, always run the reaction in a hood.*

Mercury is a commonly used liquid in the inorganic laboratory and is used in several of the experiments in this book. At 25°C, it has a vapor pressure of 1.7×10^{-3} mm Hg, which is appreciable and far above human tolerance levels if it reaches that pressure in a room. Fortunately, room ventilation usually keeps the actual level in a room far below that pressure. Mercury poisoning produces many physiological effects, including mouth sores and muscular tremors. It is primarily observed in persons who have been exposed to relatively high concentrations of mercury vapor over a period of years. This has occurred in mercury mines or in poorly ventilated laboratories where mercury was free to vaporize. For this reason, mercury should always be stored in tightly closed bottles and never allowed to spill. Spilled mercury easily hides in corners and under immovable objects, providing an undesirable source of mercury vapor. Once mercury has spilled, it is best removed by sucking it up with the vacuum flask shown in Figure I. This apparatus merely consists of a 6 mm glass tube drawn to a 1 mm tip opening, which is fitted to a trap connected to a water aspirator or vacuum pump. As much mercury as possible should be removed with this apparatus, although it is a tedious chore. The very small, invisible mercury droplets remaining should be sprinkled with powdered sulfur. Mercury reacts with sulfur to give HgS. Since the formation of HgS occurs only on the surface of the droplets and a disturbance of the droplet produces a fresh Hg surface, treatment with sulfur is at best a temporary method of reducing the vaporization of mercury. By far the best method of avoiding mercury contamination in a laboratory is not to spill it.

Chemicals present dangers besides those attributed to their toxicities. Organic solvents should all be regarded as flammable and kept far from open flames. Some chemicals produce severe burns when they come in contact with the skin. These should be handled with gloves, and an antidote should be available (e.g., $Na_2S_2O_3$ for a Br_2 burn). If a skin burn does occur, wash the area thoroughly with water and then treat it with an antidote. The disposal of some waste chemicals may be a problem. Metallic sodium waste, for example, should be reacted with anhydrous methanol to give $NaOCH_3$, which may be washed down the sink with water. Sodium reacts explosively with water and should never be placed in a waste basket. Throughout this book safety precautions are mentioned for your protection. It is almost certain that every research chemist will encounter a dangerous laboratory situation at some time; he or she must be prepared for it.

REFERENCES

P. L. Bidstrup, *Toxicity of Mercury and Its Compounds,* Elsevier, Amsterdam, Holland, 1964.

H. H. Fawcett, *J. Chem. Educ., 42,* A815, A897 (1965). A compilation of the available literature on chemical safety.

Guide for Safety in the Chemical Laboratory, Manufacturing Chemists Association, Van Nostrand Reinhold Company, New York, N.Y., 1972. Covers a broad spectrum of safety aspects, including tables of hazardous chemicals and reactions.

N. I. Sax, *Dangerous Properties of Industrial Materials,* 3rd Ed. Reinhold, New York, 1968.

S. Soloveichik, *J. Chem. Educ., 41,* 282 (1964). Effects of chemicals on great chemists.

N. V. Steere, *J. Chem. Educ., 42,* A529 (1965). Mercury vapor hazards and control measures.

N. V. Steere, ed., *Safety in the Chemical Laboratory,* Vols. I, II, & III, Chemical Education Publishing Company, Easton, Pa., 1967–1974. An excellent source of information and references on chemical safety, including a section on accident case histories.

N. V. Steere, ed., *CRC Handbook of Laboratory Safety.* 2nd Ed. Chemical Rubber Company, Cleveland, Ohio, 1971.

LABORATORY PROCEDURES

Use of Time

The efficient use of time is an asset not only to a student but especially to a researcher. Plan your experiments so that you will profitably use time that would otherwise be spent watching, e.g., a distillation, a sublimation, or a nonhazardous reaction that need not be attended. This course allows some latitude in the planning of experiments, and you should be constantly looking for opportunities to use the available time effectively.

Cleanliness

Since most of the experiments will involve the use of equipment that other students will use sometime during the course, it is absolutely essential that all equipment be left in good condition at the end of each period. Any equipment that is broken should be reported to the instructor immediately so that a replacement may be found in time for the next class.

Glassware that is difficult to clean with a detergent may be soaked in cleaning solution (~20 g $Na_2Cr_2O_7$ or $K_2Cr_2O_7$ in 80 ml concentrated H_2SO_4 and 20 ml of water) overnight. Although it works slowly at room temperature, its cleaning action may be accelerated by heating. Some pieces of glassware are more conveniently cleaned by pouring a few milliliters of concentrated H_2SO_4 into the soiled flask, followed by an equal volume of 30 per cent H_2O_2 and then by swirling the mixture. This is a very strong oxidizing mixture, which rapidly cleans almost any glassware. Needless to say, both cleaning solution and the H_2SO_4-H_2O_2 solution dissolve clothing and produce severe skin burns; they should be handled with rubber or polyethylene gloves.

Compressed Gas Cylinders

Several experiments in this book make use of gases that are commercially available in compressed gas cylinders. They come in a variety of sizes with several types of valves and regulators. For a given cylinder, the manufacturer's catalog should be consulted for the correct valve fittings. Also, the metallic content of the valves may be dictated by the corrosive properties of the gas. The facile reaction of N_2O_4 with metallic copper, for example, requires that the cylinder and valve contain very little copper. Many cylinders contain a safety valve or nut, which is designed to rupture if the pressure inside the cylinder exceeds the specifications of the cylinder. Under no circumstances should anyone tamper with the safety nut.

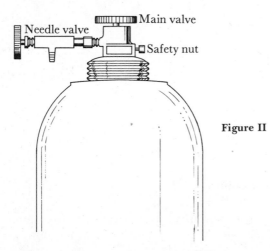

Figure II

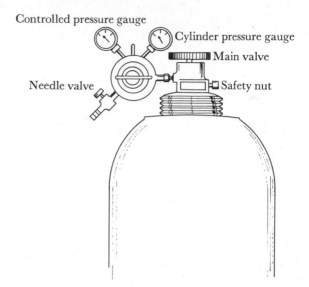

Figure III

The main valve (Figure II) on a cylinder is simply an on-off valve that allows no control of the gas flow; it should always be used with some type of control valve. A needle valve (Figure II) permits such control; but if the cylinder contains a compressed gas, the cylinder pressure will decrease as the cylinder is used and the gas flow will likewise decrease. Thus, for compounds that exist as gases (e.g., CO, N_2, Ar, BF_3, and CH_4) in the cylinder, a given flow rate cannot be maintained without continuous adjustment. Compounds that condense to form liquids under pressure exert their natural vapor pressure so long as any liquid remains in the cylinder. For these gases (e.g., CO_2, N_2O_4, NH_3, $(CH_3)_3N$, and HF), a continuous flow rate can be obtained just with a needle valve.

To achieve a constant flow rate for gases that do not condense under the pressure in the cylinder, a pressure regulator (Figure III) is required. First open the main valve; the gas pressure in the cylinder is given on the right-hand gauge. Then open the regulator valve by turning the lever *clockwise*. Finally, adjust the flow rate to the desired level by opening the needle valve. The pressure between the needle valve and the regulator valve is given on the left-hand gauge. The regulator will maintain this pressure. During the experiment, the flow can be halted by closing the needle valve, but when you are finished with the cylinder for the day, close the main valve to prevent loss of the gas in case the regulator leaks slightly. Do not empty a cylinder completely; leave approximately 25 psi (lbs/in²) (~2 atm) so that the cylinder does not become contaminated with air or other gases before it is returned to the supplier for refilling.

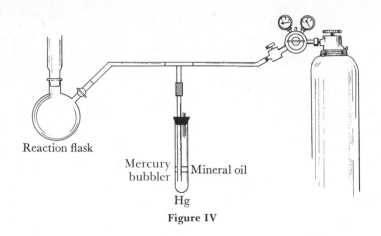

Reaction flask

Mercury   Mineral oil
bubbler

Hg

Figure IV

In several experiments, N_2 gas will be used to flush air from a re-action system, as in Figure IV. Before the reaction is begun, the N_2 flow is turned off with the stopcock on the reaction flask. This normally pro-duces a pressure build-up, which could result in the "popping" of the rubber tubing connecting the apparatus to the nitrogen cylinder. To prevent this, it is convenient to connect a mercury bubbler to the rubber tubing to act as a safety valve for any excess nitrogen pressure (Figure IV). The mercury in the bubbler is covered with a layer of mineral oil to prevent vaporization of the toxic mercury.

Utility Vacuum Line

A simple vacuum line is very convenient for carrying out routine sublimations, vacuum distillations, and the vacuum drying of products. Such a line may branch off the more sophisticated line used in Experi-ment 19 or be completely independent with its own vacuum pump. A system that we have used is shown in Figure V. Comments on the construction and use of a vacuum line are given in Experiment 19 to-gether with references to detailed accounts of vacuum line techniques. The system illustrated in Figure V is first evacuated by closing all the stopcocks opening to the air and opening those that do not. Then a Dewar flask containing liquid N_2 is placed around the cold trap. If a sublimation is to be carried out, the sublimer (Figure 11–3) may be con-nected to the vacuum through the rubber vacuum tubing. Open both the stopcock on the line and the one on the sublimer to evacuate the sublimation apparatus. When it has been completely evacuated, close the sublimer and line stopcocks and remove the sublimer. When the lower portion of the sublimation tube is warmed, the material will sublime under the static (not connected to the working pump) vacuum.

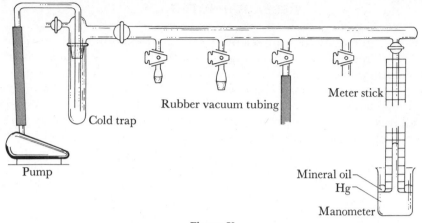

Figure V

Recrystallized materials that are still damp from the solvent may be dried by evaporating the solvent in a vacuum. The vessel (one that can withstand a vacuum, such as a round-bottom flask or tube, Figure 6–4; do not use Erlenmeyer flasks—they may implode) containing the sample is attached to the line through either a ground glass joint or the rubber tubing. It is then opened to the vacuum and allowed to dry under a dynamic vacuum with the pump operating. The evaporated solvent collects in the cold trap and should be discarded when the drying is finished. See Experiment 19, p. 195, for instructions on shutting down the vacuum line after use.

Since a fair degree of competence in glass-blowing is required to construct a vacuum line, this topic will not be covered in this book. There are, however, several books that will be helpful in improving your glass-blowing technique:

REFERENCES

R. Barbour, *Glassblowing for Laboratory Technicians,* Pergamon Press, Inc., London, 1968.

R. E. Dodd and P. L. Robinson, *Experimental Inorganic Chemistry,* Elsevier, New York, 1954, p. 90.

J. E. Hammesfahr and C. L. Stong, *Creative Glassblowing,* W. H. Freeman and Co., San Francisco, 1968.

J. D. Heldman, *Techniques of Glass Manipulation in Scientific Research,* Prentice-Hall, Englewood Cliffs, New Jersey, 1946.

A. J. B. Robertson, *Laboratory Glass-working for Scientists,* Academic Press, New York, 1957.

R. T. Sanderson, *Vacuum Manipulation of Volatile Compounds,* John Wiley and Sons, New York, 1948, Chapter 3.

R. H. Wright, *Manual of Laboratory Glass-blowing,* Chemical Publishing Co., New York, 1943.

E. L. Wheeler, *Scientific Glassblowing,* Interscience Publishers, New York, 1958.

RESEARCH NOTEBOOK

The communication of scientific facts and experimental results is an important and time-consuming duty of the scientist. Without it, however, little would be gained from the scientist's efforts. The first step in the communication chain is the accurate and detailed recording of experimental facts in a bound notebook. The purpose of this record is to allow you or someone else to learn from what you did in the experiment and to help you or them to repeat your success or avoid your failure. As you will soon learn in this course, detailed information about a synthesis or measurement is much appreciated by someone wishing to repeat your experiment. Your notebook record of your experiments should include more than enough detail to allow you or someone else to repeat the experiment successfully. It is much better to be overly detailed than to overlook observations that may be of use later.

If the experiment was a failure, as most are, your first attempt should be recorded in sufficient detail to allow you to make intelligent corrections in the procedure to increase your chance of success in the next attempt. For these reasons, your notebook should contain drawings of experimental apparatus (or a reference to a figure in this book; if the apparatus is not identical to that in the figure, alterations should be explicitly stated). It should also contain experimental observations such as color changes, temperatures of reaction mixtures, difficulties encountered, weighings, measurements, and cross references to spectra (label spectra with notebook page number on which the compound preparation is given). All of these experimental details should be recorded in the notebook at the time of the observation. Data are *not* to be first written on loose paper. This rule was not instituted by a cranky teacher; it is simply a waste of time to record observations and then recopy them into a notebook. Needless to say, if this rule is followed, your research notebook will not be a work of art in neatness, but it should be readable. Since water and acid stains will contribute to the appearance of the notebook, it is important that your records be kept with permanent ink. Mistakes should be crossed out with a single line and a brief reason stated as to why the item was lined out.

Your conclusions concerning the experiment are then to be presented. Information specifically requested at the end of each experiment should also be included at this point. Each page of the notebook should be dated to indicate the day that the experiment was performed. To facilitate referring to experiments, the first two or three pages in the notebook should be left blank for a table of contents. As you complete an experiment, its title and page should be recorded in this table of contents.

The next step in communicating results to other scientists is to prepare a formal and orderly report stressing successful techniques used to achieve the desired goal and the general conclusions resulting from the study. Such a neat, well organized report will probably not go beyond your instructor in this course, but in academic or industrial research such a record will be transmitted to other scientists in the company or university and very frequently throughout the world. They will be able to build upon your contribution.

STANDARD REFERENCES TO SYNTHESES AND TECHNIQUES IN INORGANIC CHEMISTRY

D. M. Adams and J. B. Raynor, *Advanced Practical Inorganic Chemistry,* John Wiley and Sons, New York, 1965.

G. Brauer, *Handbook of Preparative Inorganic Chemistry, Vols. I and II,* 2nd Ed., Academic Press, New York, 1963. Originally in German.

K. Burger, *Coordination Chemistry: Experimental Methods,* Chemical Rubber Company, Cleveland, Ohio, 1973.

Comprehensive Inorganic Chemistry, Pergamon Press, Oxford, England, 1973. A series of 5 volumes covering virtually all areas of inorganic chemistry.

F. A. Cotton and G. Wilkinson, *Advanced Inorganic Chemistry,* 3rd Ed., Interscience Publishers, New York, 1972. Although this text is not always mentioned at the end of each experiment, it is an excellent source of further information and references on the chemistry illustrated in many of the experiments.

D. B. Denny, ed., *Technique and Methods of Organic and Organometallic Chemistry,* M. Dekker, Inc., New York, 1969.

R. E. Dodd and P. L. Robinson, *Experimental Inorganic Chemistry,* Elsevier, London, 1957.

J. J. Eisch and R. B. King, *Organometallic Syntheses, Vol. I,* Academic Press, New York, 1965.

Gmelins Handbuch der Anorganischen Chemie, Verlag Chemie, Berlin. A comprehensive source book of inorganic chemistry.

Inorganic Syntheses, Vols. I–XV, McGraw-Hill, New York, 1939–1974. An excellent source of reliable syntheses.

W. L. Jolly, *The Synthesis and Characterization of Inorganic Compounds,* Prentice-Hall, Englewood Cliffs, New Jersey, 1970.

W. L. Jolly, *Synthetic Inorganic Chemistry,* Prentice-Hall, Englewood Cliffs, New Jersey, 1960.

W. L. Jolly, ed., *Preparative Inorganic Reactions, Vols. I–VII,* Interscience Publishers, New York, 1964–1971.

H. B. Jonassen and A. Weissberger, eds., *Technique of Inorganic Chemistry, Vols. I–VII,* Interscience Publishers, New York, 1963–1968.

H. Lux, *Anorganisch-Chemische Experimentierkunst,* 2nd Ed., Johann A. Barth Verlag, Leipzig, 1959.

W. G. Palmer, *Experimental Inorganic Chemistry,* Cambridge University Press, 1954.

G. Pass and H. Sutcliffe, *Practical Inorganic Chemistry,* 2nd Ed., Halsted Press, New York, 1974.

G. G. Schlessinger, *Inorganic Laboratory Preparations,* Chemical Publishing Co., New York, 1962.

D. F. Shriver, *The Manipulation of Air-Sensitive Compounds,* McGraw-Hill, New York, 1969.

H. F. Walton, *Inorganic Preparations,* Prentice-Hall, Englewood Cliffs, New Jersey, 1948.

A. Weissberger, *Technique of Organic Chemistry,* Interscience Publishers, New York. Several volumes, which are also valuable to an inorganic chemist.

CHECK-LIST OF PURIFICATION AND CHARACTERIZATION TECHNIQUES

In the preparation of new compounds, the actual preparative reaction is frequently the easiest aspect of the whole process of synthesizing and identifying the compound. Far more time is spent in its purification and characterization. To help organize your approach to purification and characterization of compounds, a check-list of techniques is given below. This list includes commonly used techniques but is definitely not comprehensive. Many ingenious methods have been devised for special situations; we should be constantly searching for novel techniques.

PURIFICATION TECHNIQUES

A. Recrystallization (See Expts. 1, 8, 9, 14, and 20)
B. Distillation (See Expts. 10, 17, and 19)
C. Sublimation (See Expts. 11, 14, 15, and 19)
D. Extraction (See Expt. 6)
E. Chromatography
    a) Thin-layer chromatography (See Expt. 16 and Appendix 8)
    b) Column chromatography (See Expt. 16 and Appendix 8)
    c) Ion exchange chromatography (See Expt. 7)
    d) Many other types of chromatography

CHARACTERIZATION TECHNIQUES

A. Ionic conductivity (Expt. 1 and Appendix 2)
B. Melting point
C. Molecular weight (freezing point depression, osmometry, mass spectrometry; see Expts. 14 and 19)
D. Magnetic susceptibility (See Expt. 5)
E. Infrared spectroscopy (See Expts. 1 and 14)
F. Ultraviolet-visible spectroscopy (See Expt. 2)
G. Optical rotatory dispersion and circular dichroism (See Expt. 8)
H. Nuclear magnetic resonance (See Expt. 14 and Appendix 7)
I. Mass spectrometry (See Expt. 14 and Appendix 7)
J. Electron paramagnetic resonance
K. Mössbauer spectrometry
L. Photoelectron spectroscopy

[Co(NH₃)₄CO₃]NO₃ and [Co(NH₃)₅Cl]Cl₂

Of particular importance to the development of coordination chemistry are metal complexes of the type to be synthesized and characterized in this experiment. Prior to 1950, research in this area was almost exclusively concerned with the investigation of complexes of transition metal ions with such monodentate ligands as Cl^-, Br^-, I^-, NH_3, pyridine, CN^-, and NO_2^- and bidentate ligands as ethylenediamine ($H_2NCH_2 \cdot CH_2NH_2$), oxalate ($^-O_2CCO_2^-$), glycinate ($H_2NCH_2CO_2^-$), and CO_3^{2-}. These complexes still form the basis of a vast amount of research today, despite the more recent discoveries of the ligand properties of —H, —CH_3, CO, $H_2C = CH_2$, and benzene, to mention a few.

Coordination compounds of Co(III) and Cr(III) have been of particular interest because their complexes undergo ligand exchange very slowly compared with complexes of many other transition metal ions. For example, $Ni(NH_3)_6^{2+}$ reacts virtually instantaneously with H_2O to form $Ni(OH_2)_6^{2+}$. Under the same conditions, the analogous reactions of $Co(NH_3)_6^{3+}$ and $Cr(NH_3)_6^{3+}$ occur very slowly. This difference in behavior of complexes of different metal ions has been qualitatively accounted for by ligand field theory and molecular orbital theory.

The slow reactivity of Co(III) complexes has made them suitable for extensive investigations. The structures of the octahedral Co(III) complexes which you will prepare are given below.

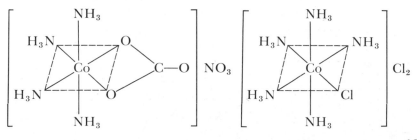

One important method of characterizing ionic substances is the determination of the ability of their solutions to conduct an electric current. Those substances whose solutions have the highest conductivity consist of the greatest number of ions. Thus, a one molar solution of $[Co(NH_3)_4CO_3]NO_3$ will have a lower conductance than a solution of $[Co(NH_3)_5Cl]Cl_2$ of the same concentration. By measuring the conductivity of a solution of a compound, it is possible to determine whether a formula unit of that compound consists of 2, 3, 4, or more ions. Although measurements will be done on water solutions of the complexes, the same information can frequently be obtained using organic solvents such as nitrobenzene or acetonitrile for ionic compounds that either are not very soluble in water or react with water.

Another very powerful method for establishing the identity of a complex is infrared spectroscopy. This technique examines the frequencies of the vibrational modes of a molecule. Thus, the infrared spectra of both of the preceding complexes exhibit absorptions at frequencies (commonly expressed in wave numbers, cm^{-1}, i.e., reciprocal wavelength $1/\lambda$) characteristic of stretching and bending modes of the NH_3 group. While the Co-N stretching modes are in principle also measurable, they sometimes occur at frequencies too low (below $650\ cm^{-1}$) to be observed in most infrared spectrophotometers. More specialized and expensive infrared instruments are, however, available for studying these vibrations. The complex $[Co(NH_3)_4CO_3]NO_3$ also exhibits absorptions characteristic of the carbonate group. Because of its coordination to the metal ion, the CO_3^{2-} group has a somewhat different spectrum in this complex than it has as the ion, as in Na_2CO_3. The spectrum should also contain absorption bands resulting from vibrational modes of the NO_3^- ion, very similar to those observed in $NaNO_3$. In contrast, the spectrum of $[Co(NH_3)_5Cl]Cl_2$ would be largely dominated by absorptions attributable to the NH_3 groups.

In general, metal-Cl stretching frequencies are lower than can be observed on usual infrared instruments, and the ionic Cl^- groups, of course, are not, in the solid state, strongly bonded to any other single atoms; thus, no absorptions are expected in the infrared spectrum to indicate their presence in the compound. While infrared spectroscopy and other instrumental methods of compound characterization are emphasized in this and in other experiments in this book, it should be stressed that a quantitative elemental analysis is an absolutely essential step in determining the composition and structure of a new compound.

The synthesis of $[Co(NH_3)_4CO_3]NO_3$ will be carried out according to the unbalanced equation,

$$Co(NO_3)_2 + NH_3(aq) + (NH_4)_2CO_3 + H_2O_2 \rightarrow$$

$$[Co(NH_3)_4CO_3]NO_3 + NH_4NO_3 + H_2O \qquad (1)$$

The $Co(NO_3)_2$ that is available commercially has the formula $Co(NO_3)_2 \cdot 6H_2O$ and very probably is a coordination compound having the ionic formulation $[Co(OH_2)_6](NO_3)_2$. Since Co(II) complexes, like those of Ni(II), react very rapidly by ligand exchange, the first step in the reaction might be expected to be $Co(OH_2)_6{}^{2+} + 4NH_3 + CO_3{}^{2-} \rightarrow Co(NH_3)_4CO_3 + 6H_2O$. This Co(II) complex could then be oxidized by the transfer of an electron to H_2O_2 to give the relatively unreactive Co(III) ion, $[Co(NH_3)_4CO_3]^+$.

The preparation of $[Co(NH_3)_5Cl]^{2+}$ is accomplished from the carbonato complex according to the following series of equations:

$$[Co(NH_3)_4CO_3]^+ + 2HCl \rightarrow [Co(NH_3)_4(OH_2)Cl]^{2+} + CO_2(g) + Cl^-$$

$$[Co(NH_3)_4(OH_2)Cl]^{2+} + NH_3(aq) \rightarrow [Co(NH_3)_5(OH_2)]^{3+} + Cl^-$$

$$[Co(NH_3)_5(OH_2)]^{3+} + 3HCl \rightarrow [Co(NH_3)_5Cl]Cl_2(s) + H_2O + 3H^+$$

On the basis of mechanistic studies of reactions of $[Co(NH_3)_4CO_3]^+$ with acids, the first reaction in the preceding sequence probably involves the following mechanism:

That O-C bond fission occurs in the intermediate has been established from ^{18}O isotopic exchange studies in several similar reactions of carbonato complexes. The subsequent steps in this preparation involve the substitution of one ligand in the coordination sphere by another. At first glance, one might expect these reactions to proceed according to S_N1 or S_N2 mechanisms, but even now there is considerable debate as to how these substitutions actually proceed. In Experiment 2, you will have an opportunity to postulate a mechanism, based on your rate data, for the reverse of the last reaction in the preceding series, the conversion of $[Co(NH_3)_5Cl]^{2+}$ to $[Co(NH_3)_5(OH_2)]^{3+}$.

EXPERIMENTAL PROCEDURE

No precautions are necessary to protect the reaction mixtures from the atmosphere. (This is required for preparations that involve reactants or products that react with moisture or oxygen in the air.)

Operations that necessitate heating of the solutions should be carried out in an efficient hood.

Carbonatotetraamminecobalt(III) Nitrate, [Co(NH$_3$)$_4$CO$_3$]NO$_3$

Dissolve 20 g (0.21 mole) of (NH$_4$)$_2$CO$_3$ in 60 ml of H$_2$O and add 60 ml of concentrated aqueous NH$_3$. While stirring, pour this solution into a solution containing 15 g (0.052 mole) of [Co(OH$_2$)$_6$](NO$_3$)$_2$ in 30 ml of H$_2$O. Then slowly add 8 ml of a 30 per cent H$_2$O$_2$ solution. (Handle H$_2$O$_2$ with rubber gloves. If the affected area is not washed immediately with water, hydrogen peroxide can cause severe skin burns.) Pour the solution into an evaporating dish and concentrate over a gas burner in a hood to 90 to 100 ml. Do not allow the solution to boil. During the evaporation time add, in small portions, 5 g (0.05 mole) of (NH$_4$)$_2$CO$_3$. Suction filter (with water aspirator; for better control of the vacuum, use a pinch clamp on the rubber tubing between the trap and filtration flask; see Figure 1-1) the hot solution and cool the filtrate in an ice water bath. Under suction, filter off the red product crystals. Wash the [Co(NH$_3$)$_4$-CO$_3$]NO$_3$ in the filtration apparatus first with a few milliliters of water (the compound is somewhat soluble) and then with a similar amount of ethanol. Calculate the yield. (Save a portion of your product for the conductance determination.)

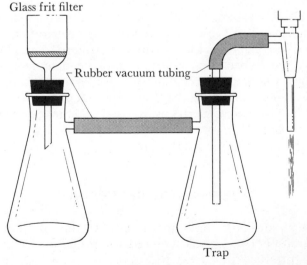

Glass frit filter

Rubber vacuum tubing

Trap

Figure 1-1

Chloropentaamminecobalt(III) Chloride, [Co(NH$_3$)$_5$Cl]Cl$_2$

Dissolve 5.0 g of $\left[\text{Co(NH}_3)_4\text{CO}_3\right]\text{NO}_3$ in 50 ml of H$_2$O and add concentrated HCl (5 to 10 ml) until all of the CO$_2$ is expelled. Neutralize with concentrated aqueous NH$_3$ and then add about 5 ml excess. Heat for 20 minutes, again avoiding boiling; $\left[\text{Co(NH}_3)_5\left(\text{OH}_2\right)\right]^{3+}$ is formed. Cool the solution slightly and add 75 ml of concentrated HCl. Reheat for 20 to 30 minutes and observe the change in color. Purple-red crystals of the product separate on cooling to room temperature. Wash the compound several times, by decantation, with small amounts of ice-cold distilled water; then filter under a water aspirator vacuum with a glass fritted funnel (medium porosity). Wash with several milliliters of ethanol. Drying in an oven at 120°C to remove solvent yields $\left[\text{Co(NH}_3)_5\text{Cl}\right]\text{Cl}_2$. Calculate the yield. (Do not discard the compound because part of it will be needed in the kinetics experiment, Experiment 2.)

Electrical Conductance of Solutions of Ionic Compounds

The determination of the number of ions constituting a given substance is largely a matter of defining conductance and then comparing conductances of known ionic substances with those of the unknown compounds. The definitions usually begin with resistance, since this is the quantity that is experimentally measured. The specific resistance, ρ, is defined as the resistance in ohms of a solution in a cell that has 1 cm^2 electrodes that are separated from each other by a distance of 1 cm. The reciprocal of ρ is the specific conductance, L. The resistance, R, of the same solution in a cell of nonstandard dimensions is obtained by multiplying ρ by a correction factor k, which depends upon the geometry of the cell. Experimentally, k is evaluated

$$R = k\rho \qquad (2)$$

by measuring R for a given solution whose ρ has been measured in a standard cell and then calculating k from the preceding expression.

Since $\rho = 1/L$, equation (2) is usually expressed in terms of measured resistance, R, and the specific conductance:

$$R = \frac{k}{L} \qquad (3)$$

The cell constant, k, is frequently obtained from equation (3) by measuring the resistance, R, of a 0.02 M KCl solution whose specific conductance at 25°C is 0.002768 ohm^{-1}. Having evaluated k for the cell used in the study, the measurement of R will allow the calculation of the specific conductance of any solution. In determining the conductance of a solution of an electrolyte, it is desirable to compare conductances under standard conditions. Thus, the molar conductance, Λ_M, is defined as the conductance of a 1 cm³ cube of solution that contains one mole (or formula weight) of solute. Since the specific conductance, L, is the conductance of a 1 cm³ cube of solution, the conductance per mole of solute may be calculated by dividing L by the number of moles present in 1 cm³ solution:

$$\Lambda_M = \frac{1000L}{M}$$

where M = molarity of the solution.

Comparisons of molar conductances with those of known ionic substances allow one to determine the number of ions present in a given salt. General ranges of Λ_M for 2, 3, 4, and 5 ion conductors at 25°C in water solvent are tabulated as follows:

Number of Ions	Λ_M
2	118–131
3	235–273
4	408–435
5	~560

Molar conductances of 2, 3, 4, and 5 ion conductors in other solvents at 25° are given in Appendix 2.

EXPERIMENTAL PROCEDURE

You will measure the conductances of aqueous solutions of your coordination compounds in a conductance cell similar to that shown in Figure 1–2. You may have made very similar measurements in your physical chemistry laboratory. To prepare the conductance cell for use, the two contact arms must be filled with mercury. (To minimize mercury spills, it is convenient to close the necks of the mercury wells with putty or a similar material.) The platinum electrodes must be coated with platinum black by connecting a 3 volt battery to the two electrode leads and adding a 1 per cent (by weight) aqueous solution of $H_2PtCl_6 \cdot 6H_2O$ to the cell. After about a minute, reverse the polarity of the leads to the battery so that platinum will be deposited on the other electrode. Continue to reverse the leads periodically until both electrodes are entirely blackened. Pour out the platinum solution and save it for future coatings of the electrodes. Wash the cell several times with distilled water and always keep the cell filled with water when it is not in use. If the cell is cared for properly, the electrodes will not need to be platinized again for months.

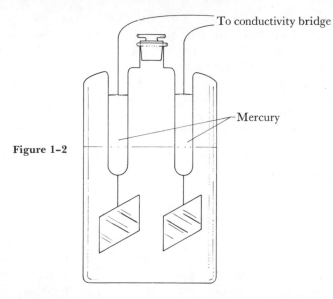

To conductivity bridge

Mercury

Figure 1–2

Read carefully the instructions provided by the manufacturer for the operation of the conductivity bridge. If you need help, ask the instructor. In making all resistance measurements, thermostat the cell containing the solution at 25°C for approximately 10 minutes before making a reading. Use distilled water in all solution preparations.

1. Prepare a 0.02M KCl aqueous solution and obtain the cell constant, k, from the resistance measured for the solution.

2. Prepare 500 ml of 0.001 M aqueous solutions of $[Co(NH_3)_4CO_3]$-NO_3 and $[Co(NH_3)_5Cl]Cl_2$ and measure their resistances. Make the measurements immediately after the solutions are prepared, since significant decomposition occurs on standing overnight.

Be certain to rinse the cell well with distilled water between measurements, and when you have completed the experiment rinse it thoroughly and leave it filled with water.

Calculate the molar conductances of the two Co(III) complexes. These measurements are fairly sensitive tests of the ionic purity of your compounds.

Infrared Spectra

There are several ways of obtaining an infrared spectrum of a sample compound: (a) as the neat (undiluted) substance, when the compound is a liquid; (b) as a gas, for volatile compounds; (c) in solution, using a suitable solvent; (d) as a mull; or (e) as a compressed solid pellet. One or more of these techniques may be used, depending on the nature of the compound under examination.

For $[Co(NH_3)_4CO_3]NO_3$ and $[Co(NH_3)_5Cl]Cl_2$, methods (a) and (b) are clearly not possible. Since the container holding the solution in method (c) must transmit infrared light, the windows in such containers (cells) frequently are made of NaCl because of its favorable optical properties and low cost. On the other hand, the use of NaCl obviates the use of water as a solvent in these cells. Other more expensive window materials are available, e.g., AgCl, CaF_2, and BaF_2, which are not appreciably soluble in water. (Transmission ranges of various materials used in infrared cells are given in Appendix 3.) Water has another disadvantage, its strong absorption of light in much of the infrared region. This problem can frequently be minimized by using D_2O, since its absorptions sometimes do not seriously interfere with the absorption of the substance in solution. For these reasons water is not often used as a solvent for infrared work.

Unless the substance is soluble in an organic solvent that does not dissolve NaCl, the sample is usually examined in the solid state, either as a mull or as a compressed pellet. If possible, it is best to measure spectra of substances in solution, since the absorption bands are much sharper and fine structure can be observed more readily. The spectra in the solid state will depend on the packing and intermolecular interactions in the crystal. These factors sometimes produce changes in frequencies and even in the number of absorptions, compared with the solution spectra.

The pellet technique requires that the desired compound be thoroughly mixed with a transparent substance such as KBr. The resulting fine powder is compressed under roughly 5000 atmospheres pressure into a thin pellet of approximately 1 cm diameter. With skill one can prepare pellets that are sufficiently transparent to be suitable for measuring their absorption spectra. This technique has been used extensively by organic chemists and has the advantage that KBr has no absorptions of its own, although H_2O is a frequent contaminant. For inorganic compounds this technique is less useful and sometimes presents serious problems. One of these is the very strong possibility of reaction with KBr. In examining the compounds prepared in this experiment, one should consider the possible exchange of Cl^- in $[Co(NH_3)_5Cl]^+$ with KBr to form $[Co(NH_3)_5Br]^{2+}$ under the conditions of high pressure required for pellet formation.

A much more convenient method of determining the spectra of solids is in a mull. It is simply a slurry or mull of the solid in the mulling agent. Frequently used mulling agents are Nujol (a brand-name for purified mineral oil that consists of C_{20}—C_{30} alkanes) and hexachlorobutadiene, both of which have characteristic absorptions in the infrared region. Their infrared spectra are shown in Figure 1–3. The selection of a mulling agent would depend upon where the absorptions of the compound occur. Sometimes it is necessary to measure the spectrum of a compound in both mulling agents to locate all of the compound's absorptions.

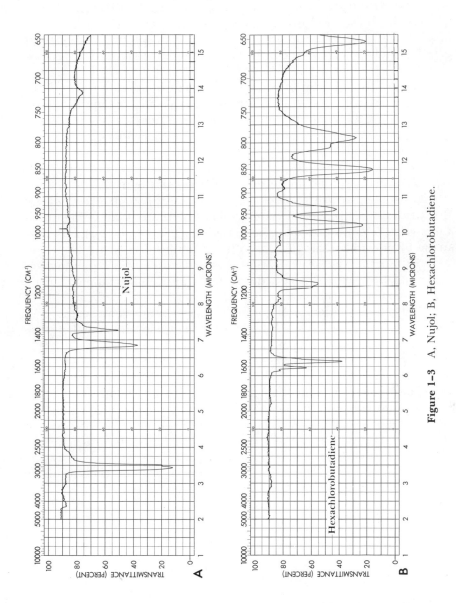

Figure 1-3 A, Nujol; B, Hexachlorobutadiene.

Preparation of a Mull.   Determine the spectrum of $[Co(NH_3)_4$-$CO_3]NO_3$ and $[Co(NH_3)_5Cl]Cl_2$ in Nujol mulls. If time permits, measure the spectrum of the former compound in a hexachlorobutadiene mull as well.

1. Put 3 to 10 mg of the complex in a polished agate mortar and grind with a pestle until the substance forms a glossy layer in the mortar (1 to 5 minutes of grinding). Small particles give the best spectra. The extent of grinding will easily make the difference between a good and a bad spectrum.

2. Add a small drop of the mulling agent by wetting the end of a spatula and touching it to the mortar. Grind this mixture until all of the sample has been taken up by the Nujol and a thick paste has formed.

3. Transfer this paste to an NaCl window with a rubber "policeman" or spatula.

4. Place a second NaCl plate window on top of the paste and *gently* rotate the plates until a thin film forms between the plates.

5. Put the plates in a holder such as is shown in Figure 1–4 and scan the infrared spectrum of the sample. The reference beam of the instrument is left open. No reliable compensation for the mulling agent can be made by inserting a sample of it in the reference beam.

6. Clean the NaCl plates by rinsing with dry acetone and, if necessary, by wiping them with a soft cloth dampened with acetone. After drying the plates in air, return them to a desiccator.

7. The preparation of Nujol mulls that produce good spectra is an art. It is entirely normal for the student to prepare several mulls before he obtains a satisfactory spectrum.

The preparation of mulls will scratch the plates, but this does not affect the spectra until they become badly gouged. A more serious problem is incomplete cleaning of the plates. Plates that are contaminated with solid compounds will obviously give absorption bands of the contaminants. When contamination is significant or the plates are badly scratched, it is necessary to polish the NaCl plates. (See references for instructions on polishing infrared window materials.)

Numerous infrared spectrophotometers are commercially available.

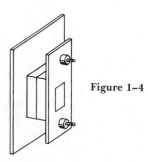

Figure 1–4

The least expensive instruments scan the most commonly used range of 4000 to 650 cm^{-1}. The principles of an infrared spectrophotometer are the same as those of other spectrophotometers, such as those operating in the ultraviolet-visible region, except that the light source emits infrared rather than visible light. (See the instruction manual and your instructor for the operation of your particular spectrophotometer.)

Before interpreting the spectrum obtained from a mull, it is important that those absorptions that result from the mulling agent be labeled directly on the spectrum. The spectrum of the mulling agent can most accurately be obtained by running it under your mull conditions on your instrument. The remaining absorption bands may then be assigned to the various vibrations of the Co(III) complexes. To make these assignments it will be necessary to consult the references at the end of the experiment.

REPORT

Include the following:
1. Percentage yields of $[Co(NH_3)_4CO_3]NO_3$ and $[Co(NH_3)_5Cl]Cl_2$.
2. Values of Λ_M for the preceding complexes and your conclusions as to the number of ions in each compound.
3. IR spectra of $[Co(NH_3)_4CO_3]NO_3$ and $[Co(NH_3)_5Cl]Cl_2$, indicating the absorptions characteristic of NH_3, CO_3, and NO_3. What are the spectral similarities and differences between these two compounds?

QUESTIONS

1. Outline a method of analyzing $[Co(NH_3)_5Cl]Cl_2$ for its percentage Cl content. (Note that the ionic Cl^- is much more reactive than that in the coordination sphere.)
2. Outline a method of analyzing $[Co(NH_3)_4CO_3]NO_3$ for NH_3 and Co content.
3. Balance equation (1) for the preparation of $[Co(NH_3)_4CO_3]NO_3$.
4. How would you experimentally establish the fission of the O—C bond, rather than the Co—O bond, in the conversion of

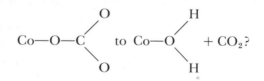

5. The conductance of an aqueous solution of $[Co(NH_3)_5Cl]Cl_2$ changes on standing overnight. Would you expect it to increase or decrease? Why?
6. In the final isolation of $[Co(NH_3)_4CO_3]NO_3$ and $[Co(NH_3)_5Cl]Cl_2$, why are the solids washed with ethanol after having first been washed with water?

7. Why are 500 ml of the 0.001M $[Co(NH_3)_4CO_3]NO_3$ and $[Co(NH_3)_5$-$Cl]Cl_2$ solutions prepared when only 50 to 100 ml are required for the conductance measurements?
8. How do $[Co(NH_3)_4CO_3]NO_3$ and $[Co(NH_3)_5CO_3]NO_3$ differ structurally? How would you experimentally distinguish between these two compounds?
9. Will the cell constant, k, change if the electrodes in a conductivity cell are bent or moved? Why?

INDEPENDENT STUDIES

A. Analyze $[Co(NH_3)_4CO_3]NO_3$ and/or $[Co(NH_3)_5Cl]Cl_2$ for their Co content.
B. Analyze $[Co(NH_3)_5Cl]Cl_2$ for its Cl^- content.
C. Prepare and characterize other $Co(NH_3)_5X^{2+}$ derivatives, where X^- = F^-, Br^-, I^-, NO_2^-, NO_3^-, CO_3^{2-}, etc. (See *Inorganic Syntheses* and G. L. Olson, *J. Chem. Educ., 46,* 508 (1969).)
D. Prepare and characterize the nitro and nitrito linkage isomers of $Co(NH_3)_5NO_2^{2+}$. (W. H. Hohman, *J. Chem. Educ., 51,* 553 (1974).)
E. Isolate and characterize *cis*-$Co(NH_3)_4(OH_2)Cl^{2+}$, which was formed as an intermediate in this experiment. (See *Inorganic Syntheses.*)
F. Exchange the protons of $[Co(NH_3)_5Cl]Cl_2$ with deuterium, using D_2O. Compare the infrared spectrum of the resulting $[Co(ND_3)_5$-$Cl]Cl_2$ with that of the proton form. (L. Sacconi, A. Sabatini, and P. Gans, *Inorg. Chem., 3,* 1772 (1964).)

REFERENCES

[Co(NH$_3$)$_4$CO$_3$]NO$_3$ and [Co(NH$_3$)$_5$Cl]Cl$_2$

K. V. Krishnamurty, G. M. Harris, and V. S. Sastri, *Chem. Rev., 70,* 171 (1970). A review of metal carbonate complexes.
K. Nakamoto, J. Fujita, S. Tanaka, and M. Kobayashi, *J. Am. Chem. Soc., 79,* 4904 (1957).
F. A. Posey and H. Taube, *J. Am. Chem. Soc., 75,* 4099 (1953).
D. B. Powell and N. Sheppard, *J. Chem. Soc.,* (London), 3108 (1956).
G. G. Schlessinger, *Inorganic Laboratory Preparations,* Chemical Publishing Co., New York, 1962, p. 219.
G. G. Schlessinger, *Inorganic Syntheses, Vol. VI,* McGraw-Hill, New York, 1960, p. 173.

Techniques

R. K. Boggess and D. A. Zatko, *J. Chem. Educ., 52,* 649 (1975). Use of conductance data for structure determination of metal complexes.
F. Daniels, J. W. Williams, P. Bender, R. A. Alberty, and C. D. Cornwell, *Experimental Physical Chemistry,* McGraw-Hill, New York, 1962. Conductance measurements.
J. R. Ferraro, *Low-Frequency Vibrations of Inorganic and Coordination Compounds,* Plenum Press, New York, 1971. Excellent source of practical and theoretical information in this area.
K. Nakamoto, *Infrared Spectra of Inorganic and Coordination Compounds,* 2nd Ed., Wiley, New York, 1970. Very useful to an inorganic chemist.

R. A. Nyquist and R. O. Kagel, *Infrared Spectra of Inorganic Compounds*, Academic Press, New York, 1971. Figures of spectra; no organometallic compounds.

W. J. Potts, Jr., *Chemical Infrared Spectroscopy, Vol. I*, Wiley, New York, 1963. Thorough account of sampling techniques.

C. N. R. Rao, *Chemical Applications of Infrared Spectroscopy*, Academic Press, New York, 1963. Sampling techniques and qualitative spectral interpretations.

General References to Coordination Compounds

F. Basolo and R. Johnson, *Coordination Chemistry*, Benjamin, New York, 1964. Elementary introduction (paperback).

K. Burger, *Coordination Chemistry*: *Experimental Methods*, Chemical Rubber Company, Cleveland, Ohio, 1973.

F. A. Cotton and G. Wilkinson, *Advanced Inorganic Chemistry*, 3rd Ed., Interscience, New York, 1972. Discussion of bonding theory of metal complexes and considerable descriptive chemistry.

M. M. Jones, *Elementary Coordination Chemistry*, Prentice-Hall, Englewood Cliffs, New Jersey, 1964. A more advanced treatment covering many aspects of the subject.

Experiment 2 ――――――――――――――

Aquation of $[Co(NH_3)_5Cl]^{2+}$

Note: $[Co(NH_3)_5Cl]Cl_2$, prepared in Experiment 1, is required for this experiment.

The large number of coordination complexes that contain H_2O as a ligand suggests that H_2O may in some cases displace more weakly coordinating ligands. Dimethyl sulfoxide (DMSO) is such a ligand. In the absence of H_2O, however, DMSO does form complexes with transition metal ions like Ni^{2+} in which coordination occurs through the oxygen atom, $Ni^{2+}-O-S{\Large\diagup}^{CH_3}_{\diagdown CH_3}$. But in H_2O, DMSO is readily displaced from the coordination sphere:

$$[Ni(DMSO)_6](ClO_4)_2 + 6H_2O \rightarrow [Ni(OH_2)_6](ClO_4)_2 + 6DMSO$$

Part of the driving force for this displacement is the large excess of the H_2O ligand-solvent. Similarly, other ligands may be replaced in other complexes. In this experiment, you will investigate the kinetics of a reaction involving the substitution of Cl^- by H_2O in acidic solution:

$$Co(NH_3)_5Cl^{2+} + H_2O \xrightarrow{H^+} Co(NH_3)_5(OH_2)^{3+} + Cl^-$$

Substitution reactions may proceed by a variety of mechanisms. One possibility is an S_N1 mechanism in which the rate determining step is the breaking of the Co—Cl bond. The resulting open coordination site in the complex is then rapidly filled by a molecule of H_2O. An S_N2 mechanism involves the attack of an H_2O molecule on the Co(III) complex to form a very short-lived seven-coordinated intermediate, which rapidly loses Cl^- to generate the product. While these two mechanisms are very different, experimentally it is impossible to differentiate between them if the reaction is conducted in H_2O solvent. The S_N1 mechanism predicts the first-order rate law,

Rate of reaction of $Co(NH_3)_5Cl^{2+} = k_1[Co(NH_3)_5Cl^{2+}]$,

where k_1 is the first-order constant expressed in units of sec^{-1}. Although the S_N2 mechanism requires the overall second-order rate law,

Rate of reaction of $Co(NH_3)_5Cl^{2+} = k_2[Co(NH_3)_5Cl^{2+}][H_2O]$,

in the case where H_2O is the solvent, it is not possible to change its concentration. Hence in this experimental situation it is impossible to determine whether the rate of reaction depends on the H_2O concentration or not. Because the nucleophile, H_2O, is in such large excess compared with the $[Co(NH_3)_5Cl^{2+}]$ concentration, and so little of the H_2O is consumed in the actual reaction, the H_2O concentration, $[H_2O]$, remains essentially constant during the progress of the reaction. Thus one would predict the experimentally observed rate law to be

$$Rate = k_{obs}[Co(NH_3)_5Cl^{2+}]$$

where $k_{obs} = k_2[H_2O]$.
Experimentally the reaction would appear to obey a first-order rate law, whereas in fact the reaction may be proceeding by an S_N2 mechanism. The problem of distinguishing S_N1 and S_N2 mechanisms for aquation reactions in water solvent has not been satisfactorily solved; hence the literature contains numerous kinetic studies of aquation reactions for which the mechanisms are not well established.

A third possible mechanism is the acid catalyzed aquation reaction. An example of this mechanism is found in the following reaction:

$$Co(NH_3)_5F^{2+} + H_2O \xrightarrow{H^+} Co(NH_3)_5(OH_2)^{3+} + F^-$$

The experimentally determined rate law is

Rate of reaction of $Co(NH_3)_5F^{2+} = k[Co(NH_3)_5F^{2+}][H^+]$

The acid catalysis apparently results from the addition of H^+ to the coordinated F^-

$$(NH_3)_5Co-F^{2+} + H^+ \underset{K}{\overset{fast}{\rightleftharpoons}} (NH_3)_5Co-FH^{3+}$$

The protonation of the coordinated F^- would be expected to weaken the Co–FH bond. One might visualize the proton as pulling off the F^- as HF, leaving an open coordination position that is quickly occupied by an H_2O molecule.

$$(NH_3)_5Co-FH^{3+} + H_2O \xrightarrow{k_3} (NH_3)_5Co(OH_2)^{3+} + HF$$

If this last step is rate-determining, this mechanism predicts the rate law

$$\text{Rate of reaction of } Co(NH_3)_5F^{2+} = k_3K\left[Co(NH_3)_5F^{2+}\right]\left[H^+\right]$$

This is the same expression that is observed experimentally. The only difference is that the experimental rate constant, k, is, in terms of this mechanism, equal to the product of equilibrium constant, K, and the rate constant for the rate-determining step, k_3:

$$k = k_3K$$

If one wishes to evaluate the rate constant k_3, some other experimental means of measuring the equilibrium constant, K, must be devised, so that k_3 may be calculated from the preceding expression.

  In this experiment you will measure the rate of aquation of $\left[Co(NH_3)_5Cl\right]^{2+}$ at different H^+ concentrations and express your rate data in the form of a rate law. You will then postulate a mechanism, which is consistent with your rate law, for the reaction. Since the starting complex and the product, $\left[Co(NH_3)_5(OH_2)\right]^{3+}$, have different extinction coefficients at 550 nm in their visible spectra, noting the change in the intensity of the absorption at that wavelength as a function of time provides a convenient means for determining the rate of the reaction. It should be recalled that the colors and associated absorption spectra of transition metal complexes usually result from electronic transitions between *d* orbitals of different energies on the metal.

EXPERIMENTAL PROCEDURE

  Although some reactions are catalyzed by light and oxygen, this reaction is not, and precautions to eliminate them need not be taken.
  In a bath set at 60°C, thermostat for at least 15 minutes two 100 ml volumetric flasks, one containing 0.10 N HNO_3 and the other 0.30 N HNO_3 solution (concentrated HNO_3 is approximately 15.9 M; see Appendix 1). To each solution add sufficient $\left[Co(NH_3)_5Cl\right]Cl_2$ (prepared in Experiment 1) to give a 1.2×10^{-2} M complex concentration. Shake the flasks until all of the complex has dissolved, and then return the flasks to the thermostatted bath. When the solution has again reached a temperature of 60°C (approximately 15 minutes), begin withdrawing samples with a pipette (use a rubber bulb, not your mouth!). At approximately 15 minute intervals withdraw 10 ml samples from each reaction mixture. Measure the visible spectrum of each sample immediately after removing

it from the thermostatted solution and record the time of withdrawal. Take 8 aliquots from each reaction solution during the course of the rate study.

There are a variety of ultraviolet-visible spectrophotometers that will allow one to follow the progress of the reaction. Most of them are double beam (reference and sample beams), and many have the capability of scanning the spectrum of a sample over the entire visible region of light. If a scanning instrument is available, the absorption spectrum of each sample should be taken over a wavelength range of approximately 350 to 650 nm. All the spectra for each kinetic run should be taken on the same piece of chart paper, and each curve should be labeled to ensure that it can later be assigned to a given solution and the time of withdrawal. While only the changes in absorption at 550 nm will be used in determining the rate law, it is of interest to note the other spectral changes that occur during the reaction. If a fixed wavelength spectrophotometer is available, the wavelength may be set at 550 nm and the absorbance at that wavelength may be recorded for each sample at a given time. While every commercially available instrument operates somewhat differently, some fundamental operations are common to all of them. Before using your particular spectrophotometer, consult the manufacturer's instructions and your instructor.

Use sample cells (see Appendix 3) of 1 cm path length for the measurements. Allow the instrument at least 15 minutes to warm-up. Before obtaining spectra of the sample, set the wavelength at 550 nm and standardize the instrument by adjusting the 100 per cent and 0 per cent transmittance controls. For the 100 per cent transmittance (zero absorbance) adjustment, place cells containing only distilled water in both the reference and sample beams. Then turn the 100 per cent adjust knob until the indicator (a meter or pen on the chart paper) reads 100 per cent T. Now completely close or block the sample beam and turn the zero adjust until the indicator reads 0 per cent T. These two controls should not be changed during the remainder of the experiment. For the spectral measurements on the samples, leave the distilled water cell in the reference beam. Empty the sample cell, rinse with acetone, and air-dry before filling with the reaction solution. When handling the cells, never touch the polished faces; fingerprints significantly reduce the transmission of light. Before installing it in the sample compartment, wipe the outside of the cell clean and dry with a soft tissue. Scan slowly the spectrum of the solution, or measure the absorbance (or transmittance) at 550 nm with a fixed wavelength instrument. Rinse the cell thoroughly with distilled water and acetone and then dry in the air or under a stream of N_2. Continue this procedure until all samples have been examined.

In kinetic studies of substitution reactions, it is usually valid to assume that the rate law will exhibit a first-order dependence on the complex concentration. Since both S_N1 and S_N2 mechanisms predict the rate law,

Rate = k_{obs}[complex], the evaluation of the measured rate constant, k_{obs}, can be made from a first-order kinetic plot of your data. If the reaction is acid catalyzed, the rate law, Rate = k[complex][H^+]^n where n = 1,2,3 etc., will hold. Since H^+ is not consumed during the reaction, the $[H^+]^n$ term is a constant in the rate law, which reduces to the expression: Rate = k_{obs}[complex], where k_{obs} = $k[H^+]^n$. Thus, k_{obs} can be obtained from a first-order kinetic plot of your data, and k_{obs} may depend upon the concentration of H^+.

From your absorbance and time data you will determine a first-order rate constant that will either be or not be dependent upon the H^+ concentration. To calculate the rate constant from your measurements you will need to know the absorbance at infinite time, A_∞. This can be calculated from the known extinction coefficient of $[Co(NH_3)_5(OH_2)]^{3+}$, which is $21.0\ cm^{-1}M^{-1}$ at 550 nm. The first-order rate constant is equal to the slope of a plot of $\ln(A-A_\infty)$ vs. time, t, where A is the absorbance of the solution at any given time, t. Justify this plot as a method of obtaining the first-order rate constant.

REPORT

Include the following:
1. Absorbance measurements or spectra taken during the kinetic run.
2. Kinetic plots of data used to determine the rate constants.
3. Justification for using a plot of $\ln(A-A_\infty)$ vs. t to obtain the first-order rate constants.
4. Rate constants and rate law.
5. Proposed mechanism.

QUESTIONS

1. Account for the fact that *trans*-[Co(NH_3)_4Cl_2]^+ undergoes substitution of one Cl^- by water at a rate that is 1000 times faster than that of $[Co(NH_3)_5Cl]^{2+}$. Both react by the same mechanism.
2. The aquation of $[Co(NH_3)_5Cl]^{2+}$ is accelerated by Ag^+. Propose a mechanism for this acceleration.
3. The optically active tris oxalato complex, $[Cr(C_2O_4)_3]^{3-}$, racemizes faster than it exchanges oxalate with free $C_2O_4^{2-}$ in solution. This latter fact was determined by using radioactive ^{14}C labeled $C_2O_4^{2-}$ in solution. Postulate a mechanism for the racemization.
4. Suggest another method, other than spectrophotometry, of determining the rate of aquation of $Co(NH_3)_5Cl^{2+}$.
5. Why was 550 nm chosen as the wavelength at which the reaction was followed?
6. Consider the reaction:

$$Co(NH_3)_5Cl^{2+} + NH_3 \rightarrow Co(NH_3)_6^{3+} + Cl^-$$

The rate law for this reaction may be written in the general form: Rate = $k[Co(NH_3)_5Cl^{2+}]^x[NH_3]^y$, where the orders, x and y, are to be determined. You carry out rate studies with $[Co(NH_3)_5Cl^{2+}] = 0.001M$ and several different NH_3 concentrations (e.g., $[NH_3] = 0.2, 0.3, 0.4$, and $0.6M$). From the results of these kinetic runs, how would you determine x and y?

7. Why was HNO_3, not HCl, used in this study? (See Experiment 1.)

INDEPENDENT STUDIES

A. Determine rates of aquation of $Co(NH_3)_5Cl^{2+}$ at two other temperatures spanning a total range of 30°C. From the results, calculate the activation parameters (E_a, $\Delta H^{\ddagger}$, $\Delta S^{\ddagger}$) for the reaction. (See a kinetics text for those calculations.)

B. Prepare $Co(NH_3)_5Br^{2+}$ and compare its rate of aquation with that of $Co(NH_3)_5Cl^{2+}$ at the same temperature. (A. W. Adamson and F. Basolo, *Acta Chem. Scand., 9,* 1261 (1955).)

C. Study the rate of hydrolysis of $Co(NH_3)_5Cl^{2+}$ in basic solution (OH⁻) to form $Co(NH_3)_5(OH)^{2+}$. (See the Adamson and Basolo reference noted above and S. C. Chan, K. Y. Hui, J. Miller, and W. S. Tsang, *J. Chem. Soc.,* 3207 (1965).)

D. Measure the equilibrium constant for the reaction:

$$Co(NH_3)_5Cl^{2+} + H_2O \rightleftarrows Co(NH_3)_5(OH_2)^{3+} + Cl^-$$

(H. Taube, *J. Amer. Chem. Soc., 82,* 524 (1960).)

E. Measure and compare the rates of 1,10-phenanthroline dissociation from its Ni(II), Co(II), and Cu(II) complexes. (G. Brumfitt, *J. Chem. Educ., 46,* 250 (1969).)

F. Prepare $Cr(OH_2)_5Br^{2+}$ and measure its rate of aquation. (I. J. Herman and A. Lifschitz, *J. Chem. Educ., 47,* 231 (1970).)

G. Determine the effect of added $Hg(NO_3)_2$ on the rate of aquation of $[Co(NH_3)_5Cl]Cl_2$. Postulate a mechanism. (F. A. Posey and H. Taube, *J. Amer. Chem. Soc., 79,* 255 (1957).)

REFERENCES

Aquation of [Co(NH₃)₅X]²⁺

A. W. Adamson and F. Basolo, *Acta Chem. Scand., 9,* 1261 (1955).

F. Basolo and R. G. Pearson, *Mechanisms of Inorganic Reactions,* 2nd Ed., John Wiley and Sons, New York, 1967, Chapter 3.

G. C. Lalor and E. A. Moelwyn-Hughes, *J. Chem. Soc.* (London), 1560 (1963).

Techniques

Any modern quantitative analysis text will contain the fundamentals of absorption spectroscopy.

General References

C. H. Bamford and C. F. H. Tipper, eds., *Comprehensive Chemical Kinetics, Vol. 7,* Elsevier Publishing Co., Amsterdam, 1972. Reviews of mechanistic studies of reactions of coordination and organometallic complexes.

F. Basolo and R. G. Pearson, *Mechanisms of Inorganic Reactions,* 2nd Ed., Wiley, New York, 1967. As comprehensive a volume as is available for the reactions of coordination compounds.

J. Burgess, D. N. Hague, R. D. W. Kemmitt, and A. McAuley, *Inorganic Reaction Mechanisms, A Specialist Periodical Report,* The Chemical Society, Burlington House, London. A series of annual volumes reviewing the literature in this area.

J. L. Burmeister and F. Basolo, *Preparative Inorganic Reactions,* W. L. Jolly, ed., Vol. 5, p. 1, Interscience, New York, 1968. The application of reaction mechanisms to the synthesis of coordination compounds.

J. O. Edwards, *Inorganic Reaction Mechanisms,* Benjamin, New York, 1963. An introduction for undergraduates (paperback).

J. C. Lockhart, *Introduction to Inorganic Reaction Mechanisms,* Van Nostrand, Princeton, New Jersey, 1966. A short text with problems (paperback).

R. G. Wilkins, *The Study of Kinetics and Mechanism of Reactions of Transition Metal Complexes,* Allyn and Bacon, Boston, 1974. A textbook including problems.

Anhydrous $CrCl_3$

Anhydrous transition metal halides, MX_n, have a strong affinity for water. Hence, anhydrous $CoCl_2$ readily reacts with H_2O to form the aquo-complex, $[Co(OH_2)_6]Cl_2$. Because of the difficulty involved in keeping $CoCl_2$ and other transition metal halides anhydrous, most of these compounds are sold commercially as the aquo-complexes. Thus, compounds known as ferric chloride, cupric chloride, or manganous chloride are actually the aquo-complexes, frequently written as $FeCl_3 \cdot 6H_2O$, $CuCl_2 \cdot 2H_2O$, and $MnCl_2 \cdot 4H_2O$.

For many purposes the aquo-complexes are satisfactory reactants, particularly for reactions conducted in water. On the other hand, there are many instances where it would be undesirable or impossible to carry out a synthesis using the hydrated transition-metal halide. Experiments 4 and 6 involve two such examples. The preparation of anhydrous metal halides may be carried out in either of two ways. First, the complex may be dehydrated by reaction with a substance that has an even greater affinity for H_2O than the metal ion. One such substance is thionyl chloride, $SOCl_2$, which reacts readily with water to form the gases SO_2 and HCl. Refluxing the hydrated transition metal halide with $SOCl_2$ results in the preparation of the anhydrous halide. In the dehydration of $FeCl_3 \cdot 6H_2O$, it proceeds as follows:

$$FeCl_3 \cdot 6H_2O + 6SOCl_2 \rightarrow FeCl_3 + 6SO_2 + 12HCl$$

Another method for preparing anhydrous halides is to use starting materials that do not contain H_2O. This is frequently done by using either the elemental metal or the metal oxide. In this experiment you will chlorinate Cr_2O_3 with CCl_4 at 650°.

$$Cr_2O_3 + 3CCl_4 \rightarrow 2CrCl_3 + 3COCl_2$$

The reaction will be carried out in a tube furnace under an inert atmosphere.

33

The anhydrous $CrCl_3$ obtained in this preparation is quite unreactive toward H_2O, but samples of the compound that are contaminated by traces of Cr(II) compounds, such as $CrCl_2$, readily yield aquo-complexes of Cr(III), e.g., $[CrCl_2(OH_2)_4]Cl$. The catalytic activity of Cr(II) complexes in this hydration reaction is almost certainly related to the fact that the ligands in Cr(II) complexes in general exchange very rapidly with other ligands. Thus, $CrCl_2$ would react very rapidly with H_2O to form $Cr(OH_2)_6{}^{2+}$. On the other hand, Cr(III) complexes undergo ligand substitution very slowly. Solid $CrCl_3$ has the structure

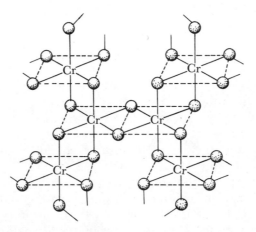

where $\bigcirc$ = Cl, in which each Cr atom is surrounded by six Cl groups, and each Cl bridges between two Cr atoms. The structure might be viewed as a series of interconnecting octahedral $Cr^{III}Cl_6$ complexes, which, like other Cr^{III} complexes, are relatively inert to ligand substitution. The Cl^- ligands are presumably labilized by the transfer of an electron from the catalytic amount of Cr(II) to the $CrCl_3$ to produce a small amount of Cr(II) in the solid, which readily reacts with H_2O to form the soluble $Cr(OH_2)_6{}^{2+}$, which could then initiate another cycle by transferring an electron to another Cr(III) in the $CrCl_3$ solid. The catalytic dissolution might be represented as follows:

$$Cr^{III}Cl_3(s) + Cr(OH_2)_6{}^{2+} \xrightarrow{\text{fast}} Cr^{II}Cl_2 + (Cl)Cr(OH_2)_5{}^{2+} \text{ (soln)}$$
inert   inert

$$Cr^{II}Cl_2 + 6OH_2 \xrightarrow{\text{fast}} Cr(OH_2)_6{}^{2+} + 2Cl^-$$
labile

While the mechanism for the dissolution of $CrCl_3$ in H_2O has not been studied in detail, our present understanding of the chemistry of Cr(II) and Cr(III) allows us to propose the very reasonable preceding mechanism.

By analogy with other redox reactions of chromium complexes, the first step of this mechanism probably involves an "inner sphere" transfer of a chlorine *atom,* as follows:

$$Cr^{III}Cl_3(s) + Cr(OH_2)_6{}^{2+} \rightarrow Cl_2Cr^{III} \cdot \overset{..}{\underset{..}{Cl}} \cdot \cdot Cr^{II}(OH_2)_5{}^{2+} + H_2O$$

$$Cl_2Cr^{II} + (Cl)Cr^{III}(OH_2)_5{}^{2+} \longleftarrow$$

This requires the initial formation of a Cl-bridged intermediate in which the $:\overset{.}{\underset{.}{Cl}}:^-$ ligand bonds to two Cr ions *via* electron pairs. Cleavage of the bridged intermediate at the wavy line leaves one electron on the Cr^{III}, thereby reducing it to Cr^{II}. The other product of the cleavage is (Cl)Cr(OH$_2$)$_5{}^{2+}$, in which an electron formally on the metal moves out to pair with the unpaired electron on the transferred Cl atom. Thus, the Cr is in the +3 oxidation state in the (Cl)Cr(OH$_2$)$_5{}^{2+}$ product. Mechanisms of redox reactions involving metal complexes are discussed in detail in the references given at the end of Experiment 2.

For this experiment, the Cr$_2$O$_3$ may either be purchased from a commercial source or be prepared in a rather spectacular reaction from (NH$_4$)$_2$Cr$_2$O$_7$. Touching a red-hot iron rod to a pile of (NH$_4$)$_2$Cr$_2$O$_7$ initiates a very exothermic reaction that generates a flurry of sparks. The equation for the display is

$$(NH_4)_2Cr_2O_7 \rightarrow Cr_2O_3 + N_2 + 4H_2O$$

In view of the high positive oxidation state of Cr and the low oxidation state of N in (NH$_4$)$_2$Cr$_2$O$_7$, it is perhaps not surprising that the compound undergoes such a vigorous self-oxidation-reduction reaction.

EXPERIMENTAL PROCEDURE

Preparation of Chromium(III) Oxide, Cr$_2$O$_3$. (This substance may already be available. See your instructor.) Place 2.5 g of (NH$_4$)$_2$Cr$_2$O$_7$ in a pile in the center of a large porcelain basin (at least 20 cm diameter). Touch a red-hot iron wire to the material until the reaction proceeds under its own exothermicity. Since hot solids will be thrown into the air, make certain that there are no flammable materials in the vicinity. The reaction should go nearly to completion, but the reactant may be reheated if this does not occur. Extract the green insoluble product with hot distilled water until the washings are colorless. Dry the Cr$_2$O$_3$ at approximately 110° C. Determine the yield.

Preparation of Chromium(III) Chloride, CrCl₃. Set up the apparatus shown in Figure 3–1. The tube furnace may be of any type. Some means of measuring the temperature must be provided. Many commercial furnaces contain built-in thermocouple temperature measuring devices. It is also possible to measure the temperature with a thermocouple connected to a potentiometer. The thermocouple may be inserted into the sample area through the open end of the tube; because of the corrosive nature of the phosgene product, the metal of the thermocouple should be protected with a glass or ceramic covering. Instead of inserting the thermocouple into the tubing, we have used a Vycor tube that was smaller than the furnace orifice; the thermocouple was then placed in the furnace outside the reaction tube. Since the proximity of the thermocouple to the heating coils will substantially affect the temperature reading, it may be necessary to raise or lower the temperature from that given in the experiment to achieve a reasonable rate of reaction.

The reaction tube must be able to withstand the 800°C temperature required. (The temperature of the tube adjacent to the heating coils is much higher than that in the sample area.) Regular Pyrex brand glass is not satisfactory; Vycor brand glass has performed well under these conditions. See Appendix 4 for properties of different types of glass.

The entire apparatus must be placed in a *hood* with the door closed, since phosgene (highly toxic; fatal in concentrations of 50 ppm) is produced as well as CrCl₃. Place 1.5 g of Cr₂O₃ in the *center* of the reaction tube. Since the temperatures at the ends of a tube furnace are much lower than at the center, the sample must be placed in the hot central area. This may be done by first fitting a piece of rubber tubing over one end of a stainless steel "Scoopula," and placing the Cr₂O₃ in the Scoopula.

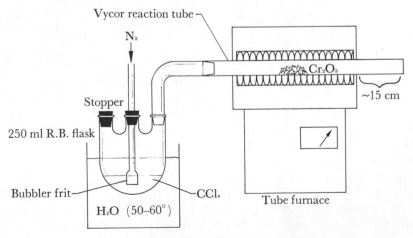

Figure 3–1

After inserting the Scoopula into the Vycor tube, empty the Cr$_2$O$_3$ into the center of the tube by rotating the rubber tubing. The Vycor tube should be wound with asbestos ribbon just inside both ends of the furnace, so that a tight fit results when the tube furnace is closed. This is necessary in order to reach the high temperatures required.

Add sufficient CCl$_4$ to cover the bubbler frit in the round-bottom flask. Heat the water bath to 50° to 60°C with a hot plate, and set the tube furnace for a temperature of approximately 800°C. Turn on the tube furnace, and begin bubbling N$_2$ slowly through the CCl$_4$. The flow of N$_2$ should be adjusted so that only gentle bubbling is observed; there should be no frothing. Too fast an N$_2$ stream will blow the Cr$_2$O$_3$ out of the Vycor tube.

When the temperature reaches about 600°, the sublimed CrCl$_3$ will begin to collect as violet flakes in the right-hand end of the tube just inside the furnace. The given reaction conditions require 1½ to 2 hours for essentially complete reaction. When no green Cr$_2$O$_3$ remains, remove the water bath, and turn off and open the tube furnace. Allow the product to cool under the N$_2$ flow before removing the reaction tube and product from the furnace. Weigh any unreacted Cr$_2$O$_3$ and also weigh the CrCl$_3$ product. The CrCl$_3$ produced in this reaction dissolves in water very slowly and is not hygroscopic. It need not be stored in a desiccator.

REPORT

Include the following:
1. Percentage yields of Cr$_2$O$_3$ and CrCl$_3$.

QUESTIONS

1. Why is an N$_2$ atmosphere maintained in the reaction tube?
2. Since the identification of CrCl$_3$ depends almost completely on its composition, outline a method of analyzing your sample for Cr and Cl.
3. What are the structures of the following compounds? FeCl$_3 \cdot 6$H$_2$O, CuCl$_2 \cdot 2$H$_2$O, MnCl$_2 \cdot 4$H$_2$O, SOCl$_2$, and COCl$_2$.
4. List equations for the synthesis of at least three other anhydrous transition metal halides.
5. How would you separate a powdered mixture of CrCl$_3$ and Cr$_2$O$_3$?
6. Why is infrared spectroscopy not a very informative method of characterizing CrCl$_3$?
7. Explain in terms of their structures why CrCl$_3 \cdot 6$H$_2$O (i.e., [trans-CrCl$_2$(OH$_2$)$_4$]Cl$\cdot 2$H$_2$O) is soluble in water but CrCl$_3$ is not. (See Experiment 7.)

INDEPENDENT STUDIES

A. Analyze your $CrCl_3$ for Cr and/or Cl.

B. Measure the magnetic susceptibility of $CrCl_3$. (See Experiment 5.)

C. Prepare an anhydrous metal chloride by dehydrating its hydrate with $SOCl_2$. (A. R. Pray, *Inorganic Syntheses, Vol. V*, McGraw-Hill, New York, 1957, p. 153.)

D. Prepare anhydrous halides such as $CrBr_3$, CrI_3, $SnCl_4$, SnI_4, $ZrCl_4$, VCl_3, $TaBr_5$, $GaCl_2$, Al_2I_6, and BiI_3 by methods described in the literature. (See *Inorganic Syntheses* and *Inorganic Laboratory Preparations* references cited at the end of this experiment.)

E. Prepare chromium nitride (CrN) in a tube furnace by the reaction: $CrCl_3 + 4NH_3 \rightarrow CrN + 3NH_4Cl$. (G. G. Schlessinger, *Inorganic Laboratory Preparations*, Chemical Publishing Co., New York, 1962, p. 22.)

REFERENCES

CrCl₃

G. B. Heisig, B. Fawkes, and R. Hedin, *Inorganic Syntheses, Vol. II*, McGraw-Hill, New York, 1946, p. 193.

A. R. Pray, *Inorganic Syntheses, Vol. V*, McGraw-Hill, New York, 1957, p. 153.

G. G. Schlessinger, *Inorganic Laboratory Preparations*, Chemical Publishing Co., New York, 1962, p. 10.

A. Vavoulis, T. E. Austin, and S. Y. Tyree, Jr., *Inorganic Syntheses, Vol. VI*, McGraw-Hill, New York, 1960, p. 129.

Techniques

R. E. Dodd and P. L. Robinson, *Experimental Inorganic Chemistry*, Elsevier, New York, 1954, p. 50. Heating methods.

W. L. Jolly, *Synthetic Inorganic Chemistry*, Prentice-Hall, Englewood Cliffs, New Jersey, 1960, p. 86. Furnace techniques.

General References to Anhydrous Transition Metal Salts

C. C. Addison and N. Logan, in *Preparative Inorganic Reactions, Vol. I*, W. L. Jolly, ed., Interscience Publishers, New York, 1964, p. 141. Preparation of anhydrous metal nitrates.

J. D. Corbett, in *Preparative Inorganic Reactions, Vol. 3*, W. L. Jolly, ed., Interscience Publishers, New York, 1966, p. 1. Methods of preparation and study of metal halides.

G. W. A. Fowles, in *Preparative Inorganic Reactions, Vol. I*, W. L. Jolly, ed., Interscience Publishers, New York, 1964, p. 121. Preparation of halide and oxyhalide compounds of the Ti, V, and Cr subgroups.

S. Y. Tyree, Jr., *Inorganic Syntheses, Vol. IV*, McGraw-Hill, New York, 1953, p. 104.

$[Cr(NH_3)_6](NO_3)_3$

Note: $CrCl_3$, prepared in Experiment 3, is required for this experiment.

In this experiment, not only will the synthesis of $[Cr(NH_3)_6](NO_3)_3$ be carried out, but techniques involved in handling liquid ammonia will be acquired.

Anhydrous liquid NH_3 has a boiling point of $-33°C$ but can be used in chemical reactions, either as a reactant or as a solvent, without any special equipment. Of course, it will vaporize at a significant rate if it is used at room temperature without external cooling. This is not, however, a serious problem if the experiments are conducted in an effective hood. In certain cases, any inconvenience created by the low boiling point of the liquid and the noxious nature of the gas is greatly exceeded by the advantageous chemical properties of the solvent.

The colorless liquid has the very unusual property of dissolving metals such as Li, Na, K, and Cs to give blue solutions. Numerous studies of the electrical conductivity of various concentrations of such solutions indicate that the metals are actually present as solvated (*i.e.,* ammoniated) metal ions and electrons:

$$M + NH_3 \rightarrow M(NH_3)_x^+ + e(NH_3)_y^-$$

Such solutions are very strong reducing agents. For example, $Ni(CN)_4^{2-}$ can be reduced with potassium metal in liquid NH_3 to give $Ni(CN)_4^{4-}$, an ion that is isoelectronic with $Ni(CO)_4$ and contains Ni in an oxidation state of zero.

$$Ni(CN)_4^{2-} + 2e(NH_3)_y^- \xrightarrow{NH_3} Ni(CN)_4^{4-} + 2yNH_3$$

In this reaction the liquid NH_3 not only provides a medium for accommodating the very reactive metallic potassium but also is a sufficiently

39

polar solvent to dissolve the reacting metal complex. The product of the reaction has been obtained as the potassium salt, $K_4[Ni(CN)_4]$.

To be sure, alkali metals do react with NH_3 but very slowly. The reaction is analogous to that of such metals with H_2O.

$$2Na + 2NH_3 \rightarrow 2NaNH_2 + H_2$$

The addition of a small quantity of a ferric, Fe(III), salt catalyzes the reaction greatly by an unknown mechanism. The ferric salt is, under the conditions of the reaction, very rapidly reduced by the ammoniated electrons to very finely divided metallic iron, which probably is the actual catalytic agent.

Sodium amide, $NaNH_2$, in NH_3 is analogous to $NaOH$ in H_2O. As OH^- is called a base in H_2O, NH_2^- is called a base in NH_3. Like water, which dissociates into hydronium and hydroxide ions,

$$2H_2O \rightleftharpoons H_3O^+ + OH^- \qquad K_w = 1.0 \times 10^{-14} \text{ at } 25°C.$$

NH_3 dissociates in a similar manner but to a significantly lesser extent:

$$2NH_3 \rightleftharpoons NH_4^+ + NH_2^- \qquad K_{NH_3} = 1.9 \times 10^{-33}, \text{ at } -50°C.$$

Thus, basic NH_3 solutions are those that contain a higher concentration of NH_2^- than of NH_4^+. An acidic solution is one that contains a relatively high concentration of NH_4^+. Needless to say, the NH_4^+ ion is not a very strong proton donor compared to H_3O^+ in H_2O, and for reactions requiring a strong acid or proton donor, liquid NH_3 is a poor choice. On the other hand, a reaction that requires a very strong base will greatly benefit from the higher basicity of NH_2^- compared with OH^-. This difference might be illustrated by comparing their reactions with acetylenes.

$$OH^- + HC{\equiv}C{-}C_6H_5 \xrightarrow{H_2O} \text{no reaction}$$

$$NH_2^- + HC{\equiv}C{-}C_6H_5 \xrightarrow{NH_3} {}^-C{\equiv}C{-}C_6H_5 + NH_3$$

From these reactions, it is clear that NH_2^- is sufficiently basic to remove H^+ but OH^- is not. The alkali metal acetylides such as $K[C{\equiv}C{-}C_6H_5]$ are relatively stable but do react with water to form KOH and the original acetylene. These acetylides have been used extensively not only in organic syntheses, but also in the formation of inorganic complexes. One inorganic application is based on the close similarity of the electronic state of the carbon atom in the $C{\equiv}N^-$ ligand to that in $^-C{\equiv}C{-}C_6H_5$. Indeed, it is possible to displace the CN^- ligand from the square planar $Ni(CN)_4^{2-}$ complex with the acetylide ion.

$$K_2[Ni(CN)_4] + 4KC\equiv C-C_6H_5 \xrightarrow{NH_3} K_2[Ni(C\equiv C-C_6H_5)_4] \downarrow + 4KCN$$

The strongly basic nature of the NH_2^- ion thus makes possible the synthesis of an unusual inorganic compound.

In this experiment you will carry out the preparation of the octahedral coordination compound, $[Cr(NH_3)_6](NO_3)_3$, according to the equations:

$$CrCl_3 + 6NH_3 \xrightarrow{NH_3} [Cr(NH_3)_6]Cl_3 \tag{1}$$

$$[Cr(NH_3)_6]Cl_3 + HNO_3 \xrightarrow{H_2O} [Cr(NH_3)_6](NO_3)_3 + HCl \tag{2}$$

If anhydrous $CrCl_3$ (prepared in Experiment 3) is simply allowed to react with liquid NH_3 as in equation (1), a major product of the reaction is $[Cr(NH_3)_5Cl]Cl_2$. The displacement of the last Cl^- proceeds very slowly; hence, the reaction is conducted in the presence of catalytic amounts of $NaNH_2$. While the mechanism of catalysis has not been investigated, it might be assumed that the greater nucleophilicity of NH_2^- as compared to NH_3 allows attack at the Cr atom with displacement of the Cl^- to give $[Cr(NH_3)_5(NH_2)]^{2+}$. This intermediate either removes a proton from NH_3 to regenerate the NH_2^- or acquires a proton when treated with acid in step (2). The product cation, $[Cr(NH_3)_6]^{3+}$, is precipitated as the relatively insoluble nitrate salt.

EXPERIMENTAL PROCEDURE

Liquid ammonia is a very volatile, irritating, and toxic material; all operations involving it *must* be conducted in a *hood*. Fortunately, its intolerably pungent odor *usually* forces the researcher from the danger area before the gas reaches fatal concentrations.

Although the boiling point of liquid NH_3 is $-33°C$, its high heat of vaporization prevents it from evaporating rapidly. Evaporation may be further minimized by cooling the reactants before dissolution in the NH_3. For many reactions, it is therefore possible to work in ordinary beakers, flasks, and other glassware. Open vessels containing liquid NH_3, of course, allow atmospheric moisture to condense into the solvent. Depending upon the sensitivity of the reactants to H_2O, protection against moisture may or may not be required. Such protection ranges from none to stoppering the flask with a drying tube to carrying out the reaction on a vacuum line. In this particular preparation the presence of small amounts of H_2O has no significant effect on the reaction. For this reason, few precautions against moisture will be taken.

Liquid NH$_3$ Solution of NaNH$_2$

The following operations *must be conducted in the hood*. It is recommended that rubber or plastic gloves be worn while handling liquid NH$_3$.

Introduce approximately 40 ml of liquid NH$_3$ into a 60 ml unsilvered Dewar test-tube or any other glass vessel of similar size (see Figure 4–1). The larger ammonia cylinders usually are fitted with a goose neck eductor tube as shown in the figure. With the cylinder tipped and the outlet of the main valve pointing upward, the eductor tube will be immersed in the liquid NH$_3$. (Smaller cylinders without eductor tubes may simply be tipped at a greater angle.)

After positioning the cylinder as shown in the figure, open the main valve (remember, this is only an on-off valve). The flow of liquid NH$_3$ is then regulated by adjusting the needle valve. Delivery of the NH$_3$ from the cylinder to the reaction vessel may be accomplished through either rubber or plastic tubing.

To the liquid NH$_3$ is added approximately 0.1 g (4.3 mmoles) of freshly cut sodium metal. (Be sure to cut off any coatings of NaOH that may be present on the sodium. Dispose of any sodium scraps by adding them to a small beaker of anhydrous CH$_3$OH, which reacts briskly to form H$_2$ and NaOCH$_3$. After the sodium has completely reacted, the solution may be washed down the drain with a flush of water. *Do not discard sodium metal in a waste basket or sink.*) The resulting blue solution is decolorized by catalyzing the reaction to NaNH$_2$ with a small crystal of Fe(NO$_3$)$_3$·9H$_2$O or other ferric salt (do not use a large amount). Stir the solution, if necessary, to discharge the blue color.

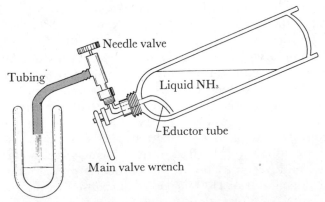

Figure 4–1

Hexaamminechromium(III) Nitrate,
$[Cr(NH_3)_6](NO_3)_3$

The dark solution, which contains finely divided iron, is treated in small portions with 2.5 g (15.8 mmoles) of finely powdered $CrCl_3$ (use mortar and pestle). Only small portions of the relatively warm $CrCl_3$ should be added in order to avoid boiling the solution over the sides of the test tube. After allowing the brown precipitate to settle, pour off the supernatant NH_3. The precipitate is transferred to an evaporating dish and allowed to dry. The solid is then quickly dissolved in 10 ml of 0.75M HCl that has been previously heated to 40°C. After suction-filtering (see Experiment 1, p. 16), 4 ml of concentrated (16M) HNO_3 is immediately added to the solution and the mixture is cooled in an ice bath. It is important that these latter steps be done as quickly as possible to avoid reaction of $[Cr(NH_3)_6]^{3+}$ with Cl^- to form $[Cr(NH_3)_5Cl]^{2+}$. The yellow $[Cr(NH_3)_6](NO_3)_3$ is suction-filtered off; it is washed with very dilute HNO_3, then with 95 per cent ethanol, and finally with diethyl ether. It is allowed to air-dry on the filter frit. Determine the yield. Since the compound slowly decomposes in light, it should be stored in a brown bottle or in a container wrapped in aluminum foil.

Save the product; you will need approximately 1 g for the determination of its magnetic susceptibility in Experiment 5.

REPORT

Include the following:
1. Percentage yield of $[Cr(NH_3)_6](NO_3)_3$.
2. Infrared spectrum (Experiment 1, p. 22) of $[Cr(NH_3)_6](NO_3)_3$ in Nujol mull, making assignments of absorptions to vibrations of the NH_3 and NO_3^- groups (optional).

QUESTIONS

1. Outline a method for quantitatively analyzing $[Cr(NH_3)_6](NO_3)_3$ for its Cr and NH_3 content.
2. Why is it recommended that gloves be worn when handling liquid ammonia? What would you do immediately if you spilled liquid ammonia on your skin?
3. If the ammonia solution of Na were not completely converted to $NaNH_2$, how would this affect the subsequent preparation of $[Cr(NH_3)_6](NO_3)_3$?
4. If a large excess of $Fe(NO_3)_3 \cdot 9H_2O$ were added to the NH_3 solution of Na, how would this affect the yield of $[Cr(NH_3)_6](NO_3)_3$ and why?
5. Postulate another mechanism, aside from that given in the intro-

duction, for the NH_2^- catalyzed conversion of $Cr(NH_3)_5Cl^{2+}$ to $Cr(NH_3)_6^{3+}$ in liquid NH_3.

6. Could $CrCl_3 \cdot 6H_2O$ (*i.e.*, $[trans\text{-}CrCl_2(OH_2)_4]Cl_2 \cdot 2H_2O$) be used in place of anhydrous $CrCl_3$ in this experiment? Why?

7. Why do N_2 cylinders require a pressure regulator (page 6) in order to maintain a constant gas pressure (or flow), while NH_3 cylinders require only a needle valve?

8. If $[Cr(NH_3)_6](NO_3)_3$ is allowed to stand in aqueous solution for a few days, the color of the solution changes. What is probably occurring?

9. Why is the precipitated $[Cr(NH_3)_6](NO_3)_3$ washed with dilute HNO_3 rather than with water? What is the purpose of the ethanol and ether washes?

10. At what stage of the $[Cr(NH_3)_6](NO_3)_3$ isolation is the metallic ion removed?

INDEPENDENT STUDIES

A. Analyze $[Cr(NH_3)_6](NO_3)_3$ for its Cr, NH_3, and/or NO_3^- content.

B. Confirm the 4-ion composition of $[Cr(NH_3)_6](NO_3)_3$ by measuring its molar conductance.

C. Determine the ultraviolet-visible spectrum of $[Cr(NH_3)_6](NO_3)_3$ in aqueous solution and assign the absorption bands to the correct electronic transitions. (H. L. Schläfer, *J. Phys. Chem.,* **69,** 2201 (1965).)

D. Prepare $[Cr(en)_3]Cl_3 \cdot 3H_2O$ and compare its ultraviolet-visible spectrum with that of $[Cr(NH_3)_6](NO_3)_3$. (R. D. Gillard and P. R. Mitchell, *Inorganic Syntheses, Vol. XIII,* McGraw-Hill, New York, 1972, p. 184.)

E. Prepare $[Cr(NH_2CH_2CH_2CH_3)_5Cl]Cl_2$ from anhydrous $CrCl_3$ and *n*-propylamine. (A. Rogers and P. J. Staples, *J. Chem. Soc.,* 6834 (1965).)

REFERENCES

$[Cr(NH_3)_6](NO_3)_3$

C. S. Garner and D. A. House, *Transition Metal Chemistry, Vol. 6,* R. L. Carlin, ed., Marcel Dekker, New York, 1970, p. 59. A thorough review of amine complexes of Cr(III).

K. W. Greenlee and A. L. Henne, *Inorganic Syntheses, Vol. II,* McGraw-Hill, New York, 1946, p. 128.

A. L. Oppegard and J. C. Bailar, Jr., *Inorganic Syntheses, Vol. III,* McGraw-Hill, New York, 1950, p. 153.

Techniques

L. F. Audrieth and J. Kleinberg, *Non-aqueous Solvents,* John Wiley, New York, 1953. Techniques of handling and applications of non-aqueous solvents.

For techniques of infrared spectroscopy, see Experiment 1, p. 19.

General References to Non-aqueous Solvents

W. L. Jolly in *Progress in Inorganic Chemistry, Vol. I,* F. A. Cotton, ed., Interscience, New York, 1959, p. 235. Physical and chemical properties of liquid ammonia solutions of metals.

H. H. Sisler, *Chemistry in Non-aqueous Solvents,* Reinhold Publishing Corp., New York, 1961. An introduction for the non-specialist (paperback).

T. C. Waddington, ed., *Nonaqueous Solvent Systems,* Academic Press, New York, 1965. A series of chapters written by specialists on such solvents as NH$_3$, HF, H$_2$SO$_4$, SO$_2$, BrF$_3$, and molten salts.

T. C. Waddington, *Non-aqueous Solvents,* Thomas Nelson and Sons, London, 1969. A teaching monograph with problems.

R. A. Zingaro, *Nonaqueous Solvents,* D. C. Heath, Lexington, Mass., 1968. A teaching monograph with problems.

For coordination chemistry references, see Experiment 1, p. 25.

Magnetic Susceptibility

Note: $[Cr(NH_3)_6](NO_3)_3$, prepared in Experiment 4, may be used in this experiment. The procedures given in this experiment may also be used in the determination of the magnetic susceptibility of any other paramagnetic or diamagnetic substance.

The widespread use of the valence bond theory of chemical bonding in transition metal complexes stemmed largely from its ability to account for and predict magnetic properties of a wide variety of coordination compounds. Although it has now been largely displaced by ligand field and molecular orbital theories, it provided a strong impetus for more extensive and detailed investigations of these properties. This theory accounts for the magnetism of complexes in terms of the number of unpaired electrons and their associated magnetic or spin properties. Those complexes that contain unpaired electrons are attracted into a magnetic field and are said to be paramagnetic, while those with no unpaired electrons are repelled by such a field and are called diamagnetic. Diamagnetism results from the fact that electrons in any material will move so as to generate a magnetic field that will oppose an applied field. Thus, any metal complex will possess some diamagnetism, although the overall character of the compound may be paramagnetic because of the very large paramagnetic susceptibility of an unpaired electron.

Thus, the magnitude of the interaction of a coordination compound with a magnetic field frequently allows the determination of the number of unpaired electrons in the complex. While this is usually true for complexes of the first-row transition elements, it is not always true for complexes of the second- and third-row metals. In these cases, the observed magnetic properties may reflect not only the spin properties of the electron but also that associated with the orbital motion of the electron. In this experiment, we will be studying the magnetic susceptibility of a coordination compound of a first-row transition metal, and from the result we will calculate the number of unpaired electrons in the complex.

A knowledge of its number of unpaired electrons is very helpful in characterizing a complex. Thus, the oxidation state of a metal in a complex may be readily assigned on the basis of its number of unpaired electrons. For example, copper in a given coordination compound will have 0 or 1 unpaired electron, depending on whether the oxidation state is +1 or +2. Also, the number of unpaired electrons may be helpful in assigning geometries to complexes. The geometries of 4-coordinate Ni complexes, $Ni^{II}L_4$, have been extensively studied by this method. Complexes of this formula may have either square planar or tetrahedral geometry. The d orbitals on Ni split, according to crystal field theory, in the following manner depending on the geometry of the complex:

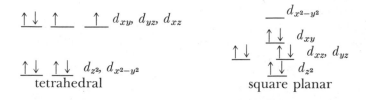

The placement of the eight d electrons of Ni^{II} into the lowest energy orbitals leaves two unpaired electrons for tetrahedral complexes and no unpaired electrons in square planar complexes. Four-coordinate Ni^{II} complexes have been found by magnetic susceptibility studies to assume either the tetrahedral or the square planar configuration, depending on the ligands bound to the metal. There are even examples of Ni^{II} complexes that in solution, exist partially in the tetrahedral and partially in the square planar form.

Finally, the number of unpaired electrons is frequently of value in indicating the presence of unusual modes of bonding. The metal carbonyl complex, $Fe_2(CO)_9$, has long been known from x-ray structural studies to have the geometry

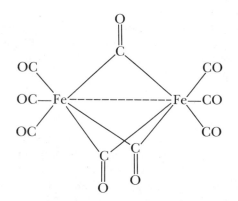

The coordination around each Fe atom is approximately octahedral. On the basis of this structure, one might hope to predict the magnetic

properties of the compound. If the ligands are formally said to be neutral, then the Fe is in the zero oxidation state and will have eight electrons. Each Fe will bond with two electrons from each of the terminal CO groups and one electron from each of the bridging CO groups. Thus, each Fe atom is surrounded by a total of 17 bonding and valence electrons. The odd number of electrons predicts that the entire molecule would have two unpaired electrons, one on each Fe. Since the compound is, in fact, diamagnetic, a reasonable bonding proposal requires that these two electrons become paired with each other. It has therefore been suggested that there is an electron-pair, covalent bond formed between the Fe atoms (dotted line in drawing) that accounts for the diamagnetism of the complex. On re-examining the structure, the relatively short Fe–Fe distance (2.46 Å) supports this suggestion. Magnetic studies have contributed greatly to our present understanding of chemical bonding in transition metal complexes.

MAGNETIC SUSCEPTIBILITY

The purpose of this experiment is to acquaint the student with a common experimental method of measuring magnetic susceptibilities. Unfortunately, there is not space in this book to provide a background in magnetism, which should be familiar to the student. It is suggested that a physics textbook be reviewed for this information. The approach here will be to make use of equations relating to magnetic fields without attempting to justify them. A rigorous treatment may be obtained in the references given at the end of this experiment.

The magnetic field inside a substance that is placed in an external magnetic field will depend not only on the magnitude of the applied field but also upon the ability of that substance to produce its own field, which will add to (paramagnetism) or subtract from (diamagnetism) the applied field. This magnetic field, B, inside the substance may then be represented as the following sum:

$$B = H_0 + 4\pi I \tag{1}$$

where H_0 is the applied field and I is the intensity of magnetization of the substance or its induced magnetic moment. The ratio I/H_0 is a measure of the susceptibility of the substance to interact with an applied field, and is called the volume susceptibility, κ. This volume susceptibility is usually converted to susceptibility per gram of substance by dividing by its density, d. It is called the specific susceptibility, χ:

$$\chi = \kappa/d. \tag{2}$$

The magnetic susceptibility per mole of substance, χ_M, is of most value for chemical applications. It is obtained by multiplying χ by the molecular weight, M, of the compound.

$$\chi_M = (\chi)(M) \tag{3}$$

The molar susceptibility, χ_M, is positive if the substance is paramagnetic and negative if diamagnetic.

Since the overall χ_M of a compound is the sum of the susceptibilities of the paramagnetic electrons in the complex and of the diamagnetic paired electrons on the metal, ligands, and ions of the compound, the susceptibility of only the unpaired electrons, χ'_M, may be obtained from the additive relationship

$$\chi_M = \chi'_M + \chi_M \text{ (metal core electrons)} + \chi_M \text{ (ligands)} + \chi_M \text{ (ions)} \tag{4}$$

In this equation, the last three terms represent the susceptibilities of the diamagnetic paired electrons of the inner core of the metal, the ligands, and the other ions (e.g., NO_3^- in this experiment), respectively. Thus, in order to determine the magnetic susceptibility of the unpaired electrons in the complex, it is necessary to correct the measured χ_M for the diamagnetism of the other groups. It is known that susceptibilities of diamagnetic groups change very little with their environment; hence, it is possible to calculate the diamagnetism of a molecule by adding together the diamagnetisms of structural components of the molecule. Some diamagnetic corrections for ligands and ions are given in Table 5–1. The corrections for transition metal ions have only been estimated (Table 5–1) and frequently only the sum $\chi'_M + \chi_M$ (metal core electrons) is evaluated.

TABLE 5–1. Diamagnetic Corrections for Ligands and Ions*

Formula	$-10^6\chi_M$, cgs	Formula	$-10^6\chi_M$, cgs
Ag^+	28	ClO_4^-	32
BF_4^-	37	F^-	9
Ba^{2+}	24	H_2O	13
Br^-	35	I^-	51
Ca^{2+}	10	K^+	15
CN^-	13	Mg^{2+}	5
NCS^-	31	NH_3	18
CO	10	NO_3^-	19
CO_3^{2-}	28	Na^+	7
$C_2H_3O_2^-$	30	OH^-	12
C_5H_5N	49	Pb^{2+}	32
C_6H_6	55	SO_4^{2-}	40
Cl^-	23	Zn^{2+}	15

*The calculated inner core diamagnetism of the first-row transition metals is approximately -13×10^{-6} cgs.

For a substance consisting of non-interacting paramagnetic centers (as in coordination compounds, where ligands insulate the metal ions from each other), the paramagnetic susceptibility should depend upon the temperature of the substance according to the Curie Law,

$$\chi'_M = C/T \qquad (5)$$

where C is the Curie constant. *If* the χ'_M of a compound obeys this law, the effective magnetic moment (in Bohr magnetons) may be calculated from the expression

$$\mu_{eff} = \left(\frac{3k\chi'_M T}{N\beta^2}\right)^{1/2} = 2.83(\chi'_M T)^{1/2} \qquad (6)$$

where k is Boltzmann's constant, β is the Bohr magneton, N is Avogaddro's number, and T is the absolute temperature.

On the other hand, the majority of compounds do not obey the Curie Law but instead follow the Curie-Weiss Law,

$$\chi'_M = C/(T + \theta) \qquad (7)$$

where θ is the Weiss constant. In this case, the expression for calculating the effective magnetic moment (in Bohr magnetons, B.M.) that is often used is

$$\mu_{eff} = 2.83\left[\chi'_M(T + \theta)\right]^{1/2} \qquad (8)$$

In order to correctly calculate μ_{eff}, it is therefore necessary to determine the temperature dependence of χ'_M. Since a temperature dependence study requires more sophisticated equipment than will be used in this experiment, we will assume that χ'_M follows the Curie Law, i.e., θ is very small. This assumption will not introduce a large error, since θ is normally only 20 to 30°.

By making the further assumption that only the spin angular momentum of the electron, and not its orbital angular momentum, contributes to χ'_M, the number of unpaired electrons, n, may be calculated from the equation

$$\mu_{eff} = \left[n(n + 2)\right]^{1/2} \qquad (9)$$

Although this latter assumption is usually valid in complexes of first-row transition metals, it is important that one realize that this assumption, as well as that pertaining to the Curie Law, will not hold in all cases.

Magnetic Susceptibilities by the Gouy Method

Experimentally, we want to determine χ_M of $[Cr(NH_3)_6](NO_3)_3$. From χ_M, the number of unpaired electrons may be calculated as indicated in the previous section. There are a number of methods of determining χ_M. Some of these involve making measurements of magnetic fields; these require expensive and delicate equipment. Ease of equipment construction and simplicity of operation have made the Gouy method the preferred method for many applications; it provides an excellent introduction to susceptibility measurements on transition metal complexes.

A schematic drawing of the experimental equipment is shown in Figure 5-1. Details of construction are given in the references at the end of the experiment. The components that we have used are given in the Notes to the Instructor (p. 225). Although the components may vary considerably, a few guidelines might be mentioned. The sensitivity of the balance must be at least 0.1 mg, although the size of the sample and magnet will also determine the quality of the balance. A silver chain is suspended from the left-hand pan of the balance through a hole in the floor of the balance case and a hole in the table. The sample tube (Figure 5-1) is attached to the silver chain with a fine copper wire and allowed to hang in the magnetic field, with the bottom of the sample tube near the center of the pole faces of the magnet. The silver chain and sample tube are enclosed in glass to prevent drafts from disturbing the weighings. The magnet may be either permanent or an electromagnet that

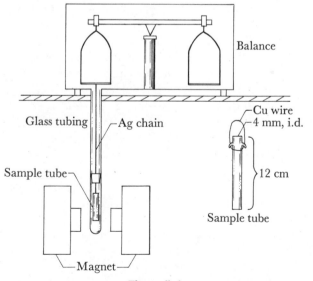

Figure 5-1

generates a magnetic field of at least 5000 gauss. Some provision must be made to remove the magnetic field from the sample, so that the sample may be weighed in the absence and presence of the magnetic field. Since an electromagnet is commonly used, the field may be simply turned off. With a permanent magnet, the magnet must be physically moved from the sample area, in order to "turn the field off."

In determining the magnetic susceptibility of a substance, one first uses a standard compound whose χ is known, for calibration of a particular Gouy apparatus. This essentially involves determining the change in weight of a given mass of standard in the presence and absence of the magnetic field. With a knowledge of the change in weight produced by a given magnetic susceptibility, χ, a measurement of the change in weight, in and out of the field, of an unknown compound will allow its χ to be calculated.

EXPERIMENTAL PROCEDURE

Scratch a horizontal line on the Gouy tube about 2 cm from the top, if this has not already been done. For the weight measurements, the tube should always be filled to this line. To remove paramagnetic impurities, clean the tube with a cleaning solution (not $K_2Cr_2O_7$–H_2SO_4; see Introduction, p. 6), thoroughly rinse with water and acetone, and dry in the oven. Do not wipe the tube with a dry towel; this gives the tube a static charge that significantly affects the weighings.

Weigh the empty tube on the chain with the field off. With the field on, weigh the tube again. Although "pure glass" is diamagnetic, paramagnetic impurities may cause the tube to be attracted by the field rather than repelled. The difference between these two weighings, "on" − "off", is δ and will be used to correct for the magnetism of the tube when the sample is weighed in the tube. It is necessary to maintain the same magnetic field for all measurements when the field is on. With electromagnets this will require a constant electric current (and therefore field). Since the current will decrease as the coils begin to heat, the current will have to be adjusted frequently. Current regulators are available that will conveniently provide this control.

To correct for the magnetism of air when it is displaced by the sample, the volume occupied by the air must be determined. This is done by weighing the tube filled to the line with water. From the weight and known density, d, of water at the existing temperature, the volume (V) may be calculated. The volume susceptibility of air is 0.029×10^{-6} per ml.

To determine the calibration constant for the apparatus, the dry tube is filled to the line with the solid standard. The largest source of error in the Gouy method is inhomogeneously packed sample tubes. To minimize this problem, the sample should be finely powdered (use a mortar and pestle) and introduced into the tube in small portions. After each addition, firmly tap the tube on a hard surface. Careful packing of the tube will require 20 to 30 minutes. Weigh the tube with the magnet off and again with the magnet on, using the same current as used previously. After each weighing of a solid sample with the field on, measure the temperature between the poles of the magnet. The difference in these two weighings ("on" − "off") is designated Δ and is a measure of the magnetic susceptibility of both the sample and the tube. Knowing the magnetic susceptibility per gram, χ, of the standard, the mass (in grams) of the standard (m), and the values of δ, Δ, and V, the calibration constant, β, may be calculated:

$$(\chi)(m) - (0.029 \times 10^{-6})V = \beta(\Delta - \delta) \qquad (10)$$

For calibration, either $HgCo(NCS)_4$ or $\left[Ni(en)_3\right]S_2O_3$, where en = $NH_2CH_2CH_2NH_2$, has proved to be very satisfactory. They may be prepared easily in high purity, are stable, are not hygroscopic, and pack very well. The susceptibility per gram, χ, of $HgCo(NCS)_4$ at 20° is $\chi_{20°} = 16.44 \times 10^{-6}$ cgs ($\pm$ 0.5 per cent). The susceptibility obeys the Curie-Weiss Law with $\theta = 10°$. The relatively high susceptibility of this compound sometimes causes the sample tube to cling to one of the poles of the magnet. This can occasionally be avoided by carefully positioning the sample tube midway between the poles. On the other hand, $\left[Ni(en)_3\right]$ S_2O_3 is rarely drawn toward a pole because of its lower susceptibility. Its value of $X_{20°}$ is 11.03×10^{-6} cgs ($\pm$1 per cent). It, too, obeys the Curie-Weiss Law with $\theta = -43°$.

After the value of β is obtained, the sample tube is cleaned and dried. The same procedure is repeated for the determination of the unknown, $\left[Cr(NH_3)_6\right](NO_3)_3$. The empty tube is weighed with the field on and off to obtain δ. Then the tube is carefully packed with finely powdered $\left[Cr(NH_3)_6\right](NO_3)_3$. The filled tube is then weighed with the field on and off to obtain Δ. These measurements will then permit the calculation of χ and the molar susceptibility, χ_M.

Summarized below are the weighings that must be made first on the standard and then on the unknown:

    A. Weight of empty tube, field off   _____g
    B. Weight of empty tube, field on   _____g
    C. Weight of tube filled to line with
       water, field off   _____g
    D. Weight of tube filled to line with
       solid, field off   _____g
    E. Weight of tube filled to line with
       solid, field on (measure temperature)   _____g

As given, these weights are related to the terms in equation (10) in the following manner:

$V = (C - A)/d$, where d is density of water (g/ml) at ambient temperature

$\delta = B - A$

$\Delta = E - D$

$m = D - A$

To determine the reproducibility of your value, the evaluation of χ for $[Cr(NH_3)_6](NO_3)_3$ should be repeated at least one more time by emptying and repacking the tube and then making the necessary weighings.

REPORT

Include the following:

1. The value of χ for $HgCo(NCS)_4$ and/or $[Ni(en)_3]S_2O_3$ at the temperature of your weighings.
2. Data necessary to calculate β, χ, χ_M, χ'_M, and μ_{eff} for $[Cr(NH_3)_6](NO_3)_3$.
3. Sample calculation of each of the above quantities, as well as the number of unpaired electrons in the complex.
4. Account for your experimentally determined number of unpaired electrons in terms of the electronic structure of the complex.

QUESTIONS

1. Account for the increase in the values of the diamagnetic corrections (Table 5-1) in going from the top to the bottom of any group in the periodic table.
2. Using the known values of χ for $HgCo(NCS)_4$ and $[Ni(en)_3]S_2O_3$, attempt the calculation of the number of unpaired electrons in each of these complexes, and interpret the results in terms of the electronic structures of the complexes.
3. An Fe(II) complex of the formula FeL_5^{2+} might have either square pyramidal or trigonal bipyramidal geometry. Would it be possible to differentiate between these geometries on the basis of a magnetic susceptibility measurement of the complex? How would the number of unpaired electrons differ in the two geometries? (Use crystal field theory to show splitting of d orbitals for the two structures.)
4. Rationalize equation (10).
5. The volume susceptibility of air is listed as 0.029×10^{-6} per ml. What causes its susceptibility to be positive?
6. If the standard were packed more compactly than the unknown, but both were packed evenly in the tube, how would this affect the calculated value of χ?

7. Why should $K_2Cr_2O_7$–H_2SO_4 cleaning solution not be used to clean Gouy tubes?
8. A student measured the effective magnetic moment of $NiCl_4{}^{2-}$ and obtained a value of 3.2 B.M. A value of 0.0 B.M. was obtained for $Ni(CN)_4{}^{2-}$. Rationalize these values in terms of the structures and crystal field splitting diagrams of these ions.
9. When the bottom of the tube is at the center of the pole faces, a Gouy tube containing $HgCo(NCS)_4$ has a certain weight in the magnetic field. If the magnet is raised so that the bottom of the tube is slightly below the center of the pole faces, will the observed weight be more or less than in the original measurement? Explain.

INDEPENDENT STUDIES

A. Determine the effective magnetic moment and number of unpaired electrons in $[Co(NH_3)_5Cl]Cl_2$, $CrCl_3$, $K_3[Fe(C_2O_4)_3] \cdot 3\ H_2O$. (R. C. Johnson, *J. Chem. Educ.*, 47, 702 (1970)), $K_3[Fe(CN)_6]$, or other compounds.
B. Using the Evans solution nmr method, determine the effective magnetic moment and number of unpaired electrons in $K_3[Fe(CN)_6]$ and other iron complexes. (J. L. Deutsch and S. M. Polling, *J. Chem. Educ.*, 46, 167 (1969); J. Loliger and R. Scheffold, *J. Chem. Educ.*, 49, 646 (1972); D. Ostfeld and I. A. Cohen, *J. Chem. Educ.*, 49, 829 (1972).)
C. Determine the electron spin resonance spectrum of $K_3[Mo(CN)_8]$ or $K_3[W(CN)_8]$. (B. R. McGarvey, *Inorg. Chem.*, 5, 476 (1966).) For general discussion of the electron spin resonance technique, see the following references: A. Abragam and B. Bleaney, *Electron Paramagnetic Resonance of Transition Ions,* Oxford University Press, London, 1970; R. S. Drago, *Physical Methods in Inorganic Chemistry,* Reinhold Publishing Corp., New York, 1965, Chapter 10; B. A. Goodman and J. B. Raynor, *Adv. Inorg. Chem. Radiochem.*, 13, 135 (1970); B. R. McGarvey, *Transition Metal Chemistry,* R. L. Carlin, ed., Marcel Dekker, New York, 1966, p. 89.

REFERENCES

Magnetochemistry—Techniques and Interpretation

T. G. Dunne, *J. Chem. Educ.*, 44, 142 (1967).
A. Earnshaw, *Introduction to Magnetochemistry,* Academic Press, London, 1968.
B. N. Figgis and J. Lewis in *Technique of Inorganic Chemistry, Vol. IV,* H. B. Jonassen and A. Weissberger, eds., Interscience Publishers, New York, 1965, p. 137.
B. N. Figgis and J. Lewis in *Modern Coordination Chemistry,* J. Lewis and R. G. Wilkins, eds., Interscience Publishers, New York, 1960, p. 400.
P. W. Selwood, *Magnetochemistry,* Interscience Publishers, New York, 1956.

CrCl$_3$(THF)$_3$

Note: Anhydrous CrCl$_3$, prepared in Experiment 3, is required for this experiment.

The history of organometallic chemistry includes what initially seem to be numerous unrelated studies of reactions of metal salts with organic molecules. One of these investigations involved the work of Hein in 1919 on the reaction of anhydrous CrCl$_3$ with the Grignard reagent C$_6$H$_5$MgBr. The initial product was formulated as "(C$_6$H$_5$)$_5$Cr·Br." Depending upon the reaction conditions and the halide ion present, compounds of the formulas (C$_6$H$_5$)$_4$CrI and (C$_6$H$_5$)$_3$CrI could also be obtained. The nature of these compounds remained unknown for many years until Zeiss and Tsutsui in 1954 developed better methods of synthesizing the compounds and gained a more complete understanding of the interactions of CrCl$_3$ with ether solvents and with C$_6$H$_5$MgBr. One complex that is undoubtedly an intermediate in these reactions results from the reaction of CrCl$_3$ with the solvent, tetrahydrofuran. This coordination compound, CrCl$_3$(THF)$_3$, is the subject of this experiment.

Anhydrous CrCl$_3$ is insoluble in ether solvents. It was noted in Experiment 3, however, that in the presence of small amounts of Cr^{II}, CrCl$_3$ dissolves readily in water to form complexes, CrCl$_2$(OH$_2$)$_4^+$, CrCl(OH$_2$)$_5^{2+}$, and Cr(OH$_2$)$_6^{3+}$, in which water acts as a ligand. If an ether solvent such as tetrahydrofuran (THF), $\begin{matrix} \text{CH}_2\text{---CH}_2 \\ | \qquad\qquad \diagdown \\ \text{CH}_2\text{---CH}_2 \diagup \end{matrix}\text{O}$, is used to dissolve CrCl$_3$, one might assume that a reaction analogous to that in water would occur. Hence, the reaction of CrCl$_3$ with THF in the presence of a catalytic amount of powdered zinc metal proceeds according to the equation:

$$\text{CrCl}_3 + 3\text{THF} \xrightarrow{\text{Zn}} \text{CrCl}_3(\text{THF})_3$$

The role of the zinc is to reduce some of the Cr^{III} in $CrCl_3$ to Cr^{II}. The Cr^{II} catalyzes the dissolution of $CrCl_3$ in THF with the resulting complex formation. An x-ray structural investigation of the analogous bromide complex shows that the tetrahydrofuran molecules are bonded through the ether oxygen atom, $Cr-O\underset{CH_2-CH_2}{\overset{CH_2-CH_2}{\big<}}$, and that the ligands have a *trans* arrangement in the octahedral complex.

Although the structure of the chloride complex, $CrCl_3(THF)_3$, has not been determined, it presumably is the same. Because THF is not a very strong donor ligand, $CrCl_3(THF)_3$ reacts with water with displacement of THF by the more strongly coordinating H_2O.

Since $CrCl_3(THF)_3$ is soluble in tetrahydrofuran, its reaction with C_6H_5MgBr proceeds smoothly to give $Cr(C_6H_5)_3(THF)_3$.

$$CrCl_3(THF)_3 + 3C_6H_5MgBr \rightarrow Cr(C_6H_5)_3(THF)_3 + 3MgBrCl$$

This, too, is probably an octahedral complex containing

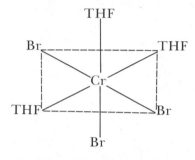

bonds. Under high vacuum the complex loses all three molecules of THF to give a black pyrophoric material, which on treatment with H_2O in the absence of air yields a mixture of π-arene complexes as follows:

$$Cr(C_6H_5)_3(THF)_3 \xrightarrow[\text{vacuum}]{- 3THF} \text{pyrophoric material}$$
$$\downarrow H_2O$$

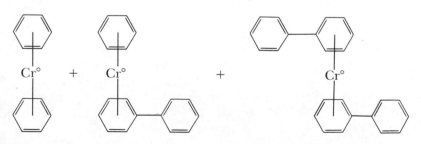

These are "sandwich" compounds, analogous to ferrocene, and have been well characterized. In the presence of air, or other oxidizing agents, they are oxidized to the cationic complexes. For example,

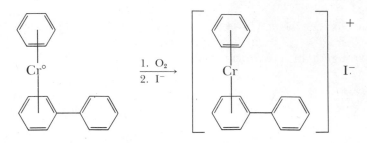

The original compound of Hein, which was formulated as $(C_6H_5)_3CrI$, has, as a result of these studies, now been established as the ionic π-arene complex shown above. The chemistry of π-arene complexes of transition metals has expanded greatly since the pioneering work on chromium. Major contributions in this general area of organometallic chemistry were made by E. O. Fischer and G. Wilkinson, who were awarded the Nobel Prize for Chemistry in 1973.

EXPERIMENTAL PROCEDURE

Tetrahydrofuran (THF) is commercially available in various degrees of purity. Because of the hygroscopic nature of THF, it is frequently contaminated with water. In this preparation, the THF solvent must be absolutely dry. Before drying the solvent, however, it is necessary to check for the presence of peroxides by adding about 1 ml of the THF to an equal volume of glacial acetic acid containing approximately 0.10 g of NaI or KI. A yellow color indicates a low concentration and brown indicates a high concentration of peroxides. Solvent giving a brown test should be discarded by flushing down the drain with excess water. Peroxides present in smaller concentrations *must* be removed in order to avoid a possible explosion. This can be done by refluxing a 0.5% suspension of cuprous chloride (CuCl) in the THF for 30 minutes, followed by distillation.

When the solvent is peroxide-free, it is first pre-dried over NaOH pellets for one day. It is then decanted (or filtered, if necessary) into another container to which are added several grams of metallic Na freshly cut into small pieces. (See note on Na disposal, Experiment 4, p. 42.) It is allowed to dry overnight. Since H_2 is evolved in the reaction of Na with H_2O, the container cannot be closed tightly, yet it must not be open to atmospheric moisture. This is accomplished by using either a drying tube or a bubbler (Figure 6–1). Drierite (CaSO₄) is a convenient

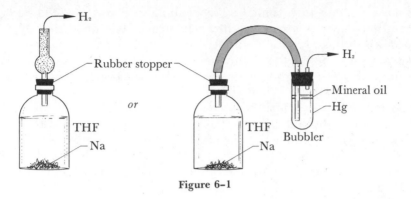

Figure 6–1

desiccant (see Appendix 6) for the drying tube, and mercury covered with mineral oil may be used in the bubbler. Preliminary drying over NaOH is not necessary for high quality commercial THF. The quality can be determined by noting the vigor of the reaction of THF with a small piece of metallic Na. Tetrahydrofuran that reacts vigorously should first be dried over NaOH.

CrCl$_3$(THF)$_3$

The reaction vessel to be used in this experiment is a Soxhlet extractor (Figure 6–2). It is normally employed in the separation of cer-

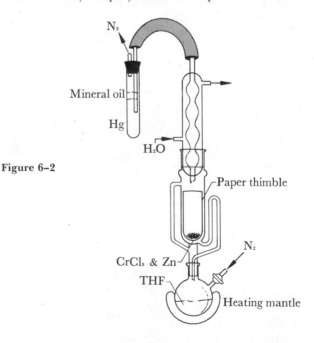

Figure 6–2

tain components from a mixture by continuous extraction. This experiment illustrates its use in synthetic chemistry. The quantities of THF specified in this preparation are those required for the small commercially available Soxhlet extractor with a 34/45 standard taper joint at the bottom of the condenser. Into the paper extraction thimble is placed 1.0 g of finely powdered anhydrous $CrCl_3$ (use mortar and pestle), which is mixed with approximately 0.1 g of powdered zinc. After adding 100 ml of sodium-dried THF to a 250 ml round-bottom flask with a stopcock side-arm, assemble the apparatus as shown in Figure 6–2 (be sure to lightly grease the ground glass joints), and flush the system with a stream of nitrogen gas through the side-arm for about 5 minutes. (See Introduction for handling of compressed gas cylinders.) Close the stopcock to stop the nitrogen flow and close the valve on the cylinder. Turn on the water in the condenser and heat the THF (b.p., 66°C) to reflux. The THF will vaporize, condense in the condenser, drop into the thimble, and react with the $CrCl_3$. When the THF level in the thimble and the right-hand side-arm has reached the top of the side-arm, the solution will be siphoned through the side-arm into the round-bottom flask. The soluble $CrCl_3(THF)_3$ will collect in the round-bottom flask and the $CrCl_3$ will, by repeated extractions, be largely converted to the desired product. After 2½ hours of reaction, the heating mantle is turned off, and the N_2 flow is immediately turned on to prevent the mercury in the bubbler from being drawn back into the reaction mixture. When the round-bottom flask has cooled to near room temperature, remove it, with the N_2 still flowing to prevent air from entering the flask. Stopper the flask and turn off the N_2. Remove the N_2 connection from the flask. Under a water aspirator vacuum, reduce the volume of the solution

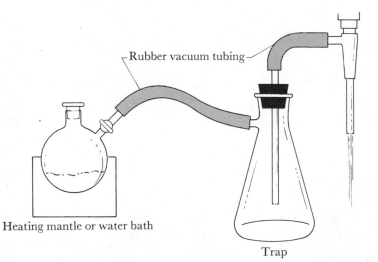

Rubber vacuum tubing

Heating mantle or water bath

Trap

Figure 6–3

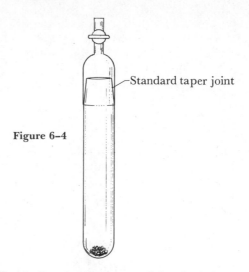

Standard taper joint

Figure 6–4

(Figure 6–3) to approximately 10 ml. Gentle warming of the flask in a heating mantle or a beaker of warm water will significantly shorten the evaporation time. The resulting precipitate is rapidly filtered on a medium frit (Figure 1–1), placed in a tube (Figure 6–4), and dried on the vacuum line (see Introduction) under a dynamic vacuum for about 1 hour. The product, CrCl$_3$(THF)$_3$, will noticeably react with water in the air within 30 minutes. The rate of this reaction depends on the atmospheric humidity. Determine the yield and store in a desiccator.

Its high solubility in organic solvents permits one to obtain the infrared spectrum of CrCl$_3$(THF)$_3$ in CHCl$_3$ solvent. (See Experiment 1, p. 19, and Experiment 14, p. 136, for infrared techniques.)

REPORT

Include the following:
1. Yield of CrCl$_3$(THF)$_3$.
2. Infrared spectrum and interpretation.

QUESTIONS

1. What is the purpose of using Hg covered by mineral oil in the bubbler?
2. Devise a scheme for analyzing CrCl$_3$(THF)$_3$ for its Cr and Cl content.
3. Predict the magnetic properties of the π-complexes Cr(C$_6$H$_6$)$_2$ and Cr(C$_6$H$_6$)$_2$$^+$.
4. What is the probable structure of the compound that Hein originally formulated as (C$_6$H$_5$)$_4$CrI?
5. Draw the probable structure of ScCl$_3$(THF)$_3$. (J. L. Atwood and K. D. Smith, *J. Chem. Soc., Dalton Trans.*, 921 (1974).)

6. Using balanced equations, explain how powdered Zn catalyzes the dissolution of $CrCl_3$ in THF.
7. Is $CrCl_3(THF)_3$ diamagnetic or paramagnetic? Explain.
8. The $CrCl_3(THF)_3$ complex reacts rapidly with excess water. What are the most probable products? (See Experiment 7.)

INDEPENDENT STUDIES

A. Analyze $CrCl_3(THF)_3$ for its percentage Cr and/or Cl content.
B. Prepare $CrCl_3(pyridine)_3$ from anhydrous $CrCl_3$ and pyridine. (J. C. Taft and M. M. Jones, *Inorganic Syntheses, Vol. VII*, McGraw-Hill, New York, 1963, p. 132.)
C. Prepare $CrCl_3(dma)_3$, where dma = dimethylacetamide, from $CrCl_3 \cdot 6H_2O$ and dimethylacetamide. (C. L. Rollinson and R. C. White, *Inorg. Chem., 1*, 281 (1962).)
D. Determine the effective magnetic moment and number of unpaired electrons in $CrCl_3(THF)_3$.

REFERENCES

$CrCl_3(THF)_3$

J. P. Collman and E. T. Kittleman, *Inorganic Syntheses, Vol. VIII*, McGraw-Hill, New York, 1966, p. 150.
A. J. Gordon and R. A. Ford, *The Chemist's Companion,* John Wiley and Sons, New York, 1972, p. 437. Methods of testing and purifying ether solvents of peroxides.
W. Herwig and H. H. Zeiss, *J. Org. Chem.,* 23, 1404 (1958).
R. J. Kern, *J. Inorg. Nucl. Chem., 24,* 1105 (1962); G. W. A. Fowles, D. A. Rice, and R. A. Walton, *J. Inorg. Nucl. Chem., 31,* 3119 (1969); L. S. Benner and C. A. Root, *Inorg. Chem., 11,* 652 (1972). THF complexes of other transition metal halides.

π-Benzene Complexes of Transition Metals

H. Zeiss, P. J. Wheatley, H. J. S. Winkler, *Benzenoid-Metal Complexes,* Ronald Press Co., New York, 1966. A survey of preparations, properties, and structures of π-arene complexes of the transition metals.
H. Zeiss in *Organometallic Chemistry,* H. Zeiss, ed., Reinhold Publishing Corp., New York, 1960, p. 380. An earlier account of π-arene complexes of transition metals with some emphasis on historical aspects.

Ion Exchange Separation of Ionic Complexes

The separation of ions by ion exchange chromatography has touched almost every area of chemistry. In inorganic chemistry, the separation of the +2 ions of the transition metals has become a routine technique for the isolation of high purity salts. Of particular note is the separation of the +3 ions of the different lanthanides by ion exchange chromatography. Their separation has not been achieved by any other means. Another outstanding success of the technique has been the quantitative separation of amino acids from mixtures produced by the degradation of proteins.

While the affinity of a particular ion for an ion exchange resin is governed by many factors, the magnitude of the charge on the ion is one of the most important. On this basis it is expected that the affinities of the following cations for anionic sites in a resin will increase in the order:

$$CrCl_2(OH_2)_4{}^+ < CrCl(OH_2)_5{}^{2+} < Cr(OH_2)_6{}^{3+}$$

It is these differing affinities for a resin that allow the separation of these cationic complexes in this experiment.

Most commercially available ion exchange resins are insoluble polymers resulting from the copolymerization of styrene and various isomers of divinylbenzene.

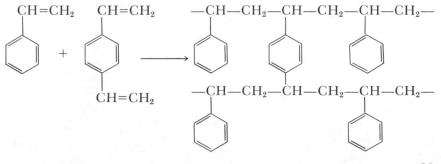

The relative amount of divinylbenzene that is used in the polymerization greatly affects the solubility and physical properties as well as the ion exchange properties of the eventual resin. This results from the extent of crosslinking of styrene chains by divinylbenzene. In general, although not always, increased crosslinking reduces the sizes of the pores into which the ions must diffuse in order to reach ionic sites in the resin. Resins that contain high proportions of divinylbenzene are, for this reason, more selective toward small ions. Highly crosslinked resins also swell relatively little when they come in contact with water. The resin (Dowex 50W-X8) to be used in this experiment contains a moderate amount, 8 per cent, of divinylbenzene, which is indicated by its "X" number. Although the original commercial resins were of dark brown color, many resins are now available in a yellow or "white" form. These light-colored resins are convenient for observing the progress of colored ions on an ion exchange chromatography column.

Into the polystyrene polymer are introduced the desired ionic groups. The cation exchange resin that will be used in this experiment contains $-SO_3H$ groups and may be prepared simply by the reaction of polystyrene with H_2SO_4.

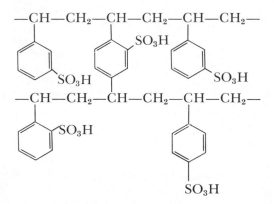

Cation exchange resins that contain $-COOH$ or $-PO_3H_2$ groups are also available. Anion exchange resins usually contain the ammonium halide group, $-N(CH_3)_3{}^+Cl^-$.

The resin used in this experiment is obtained in the hydrogen ion (H^+) form, which is represented as $RSO_3{}^-H^+$. Other cations, M^+, that may be added to the resin will also have an affinity for the anionic sites and will, to a greater or lesser extent depending upon the nature of the cation, displace the H^+ ions. In general, the equilibrium

$$RSO_3{}^-H^+ + M^+ \rightleftarrows RSO_3{}^-M^+ + H^+$$

will be established. The value of the equilibrium constant for this interaction will be unique for any given M^+, and the position of the equilibrium will depend upon the relative concentrations of M^+ and H^+ in solution. When the solution has a low H^+ concentration, M^+ will associate with the SO_3^- groups to the greatest extent. By increasing the H^+ concentration it is possible to displace the ion from the resin. For two different cations, M_1^+ and M_2^+, displacement of the cation, M_1^+, which has the lower affinity for the resin, will occur at relatively low H^+ concentration while higher acid concentrations will be required to displace the more tightly bound M_2^+.

As previously mentioned, we expect $CrCl_2(OH_2)_4^+$ to be less strongly associated with the resin than $Cr(OH_2)_6^{3+}$. The separation of the three ions in this experiment is based on the greater affinity of the more highly charged cations for the resin and on the higher H^+ concentrations which are demanded for their removal. If a solution of low acidity (2×10^{-3} M $HClO_4$) containing $CrCl_2(OH_2)_4^+$, $CrCl(OH_2)_5^{2+}$, and $Cr(OH_2)_6^{3+}$ is placed on a column of $RSO_3^-H^+$ resin (Figure 7–1), all of the complex ions will strongly adhere. None of them may be eluted from the resin by pouring 2×10^{-3} M $HClO_4$ onto the column. On increasing the H^+ concentration to 0.1 M $HClO_4$, the equilibrium will be sufficiently displaced to liberate the most weakly held cation, $CrCl_2(OH_2)_4^+$, which will be eluted from the column. It is necessary to increase the H^+ concentration to 1.0 M $HClO_4$ to cause $CrCl(OH_2)_5^{2+}$ to move down the resin. Finally, 3.0 M $HClO_4$ is required to displace the very strongly bound $Cr(OH_2)_6^{3+}$ from the resin.

The Cr(III) ions that will be separated are derived from the compound that is usually designated as $CrCl_3 \cdot 6H_2O$. An x-ray structural investigation of it shows that this complex is actually $[\textit{trans}-CrCl_2(OH_2)_4]$ $Cl \cdot 2H_2O$. In the solid state, it contains discrete $\textit{trans}-CrCl_2(OH_2)_4^+$, Cl^-, and H_2O units. The uncoordinated molecules of H_2O are hydrogen-bonded to each other and to the H_2O molecules in the complex. In aqueous solution $\textit{trans}-CrCl_2(OH)_4^+$ undergoes aquation to form $CrCl(OH_2)_5^{2+}$. In 1.0 M HCl solution at 25°C, this reaction is known to

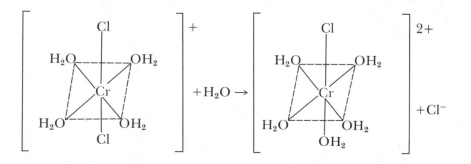

proceed with a half-life of approximately 2½ hours. The product undergoes further aquation under the same conditions with a half-life of approximately 700 hours to give hexaaquochromium(III). At lower acid concentrations, the reaction occurs considerably faster. Because of this

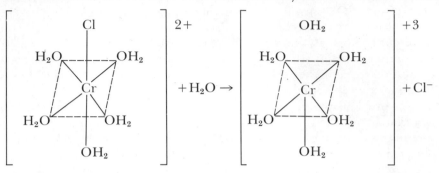

last reaction, the preparation of pure $CrCl(OH_2)_5^{2+}$ is quite difficult. The difference in affinity of these three complexes for a cation exchange resin, however, has permitted the isolation of pure solutions of $CrCl(OH_2)_5^{2+}$.

In this experiment a solution of *trans*-$CrCl_2(OH_2)_4^+$ will be converted to $CrCl(OH_2)_5^{2+}$ and then to $Cr(OH_2)_6^{3+}$. By ion exchange chromatography, pure solutions of each complex will be obtained, and the complexes will be characterized by their visible spectra. From a knowledge of the spectra of these ions, it will then be possible to isolate and identify the ions that are present in an aged aqueous solution of *trans*-$CrCl_2(OH_2)_4^+$.

EXPERIMENTAL PROCEDURE

Preparation of Ion Exchange Column

A column of the approximate dimensions shown in Figure 7–1 may be conveniently used in this experiment. Fill the column three-quarters full with distilled water. Push a small plug of cotton or glass wool to the bottom of the column with a glass rod. Pour a water slurry of Dowex 50W-X8 (50–100 mesh, H+ form) cation exchange resin into the column until a final resin height of approximately 30 cm is achieved. Allow water to pass through the resin until the effluent is colorless. Then lower the water level so that it coincides with the top of the resin. If the water level drops below that of the resin, channels will develop. Because channeling reduces the separation efficiency of a resin, never allow the resin to go dry.

Prepare 200 ml solutions each of 0.1 M, 1.0 M, and 3.0 M $HClO_4$ (see Appendix 1). These will be used to elute the desired complexes.

Prepare 100 ml of 0.35 M $CrCl_2(OH_2)_4^+$ by dissolving commercial $CrCl_3 \cdot 6H_2O$ in 100 ml of 0.002 M aqueous $HClO_4$. Portions of this solu-

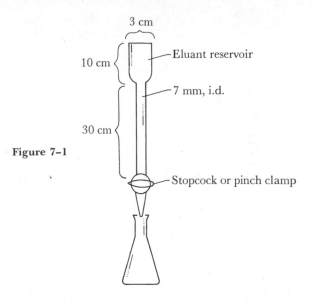

Figure 7-1

tion will be used throughout the experiment. This solution should be prepared immediately before separating $CrCl_2(OH_2)_4^+$.

trans-Dichlorotetraaquochromium(III) Ion, *trans*-$CrCl_2(OH_2)_4^+$

Put 5 ml of the 0.35 M $CrCl_2(OH_2)_4^+$ solution on the cation exchange column previously prepared, and drain until the solution and resin levels are the same. Add 0.1 M $HClO_4$ to the column and elute $CrCl_2(OH_2)_4^+$, using a flow rate of approximately 1 drop per second. Approximately 5 ml of the $CrCl_2(OH_2)_4^+$ solution will be required for the recording of its visible spectrum. If the solution is too dilute, the spectrum will not be of value. For this reason, collect only the portion of the eluate with a relatively intense color. Record the spectrum of the effluent *trans*-$CrCl_2(OH_2)_4^+$ solution on a visible spectrophotometer in the wavelength range from 350 to 650 nm, using glass cells of 1 cm path length (see Experiment 2, p. 29, for comments on UV-visible spectrophotometers). Discard the resin, including any complexes with a higher charge that remain on the resin, and prepare a new column.

Chloropentaaquochromium(III) Ion, $CrCl(OH_2)_5^{2+}$

$CrCl_2(OH_2)_4^+$ may be largely converted to $CrCl(OH_2)_5^{2+}$ by warming in aqueous solution. Swirl an Erlenmeyer flask containing 10 ml of the 0.35 M $CrCl_2(OH_2)_4^+$ solution in a 50° to 55°C water bath or under a hot water tap for $1\frac{1}{2}$ minutes. Immediately add 10 ml of distilled water and pour the entire solution on a fresh column. After draining the solution to the resin level, wash the column with 0.1 M $HClO_4$ until the unreacted $CrCl_2(OH_2)_4^+$ has been eluted. The desired complex, $CrCl(OH_2)_5^{2+}$, is then eluted with 1.0 M $HClO_4$. Again, approximately 5 ml of a relatively intensely colored fraction of the complex is collected. Measure the visible spectrum of the solution. Discard the resin and prepare a new column.

Hexaaquochromium(III) Ion, $Cr(OH_2)_6^{3+}$

Dilute 10 ml of the 0.35 M $CrCl_2(OH_2)_4^+$ solution with 10 ml of distilled water and boil for 5 minutes. To the resulting solution of $Cr(OH_2)_6^{3+}$, add an additional 10 ml of water and cool to room temperature. Add all of this solution to the column and drain it until the solution level reaches that of the resin. First rinse the column with 1.0 M $HClO_4$ to remove any unreacted $CrCl_2(OH_2)_4^+$ or $CrCl(OH_2)_5^{2+}$. Then elute the complex, $Cr(OH_2)_6^{3+}$, with 3.0 M $HClO_4$. Collect approximately 5 ml of an intensely colored portion of the eluted $Cr(OH_2)_6^{3+}$ solution and record its visible spectrum.

Separation of $CrCl_2(OH_2)_4^+$, $CrCl(OH_2)_5^{2+}$, and $Cr(OH_2)_6^{3+}$ from a Mixture

While these ions were being purified and characterized by their visible spectra, the $CrCl_2(OH_2)_4^+$ in the original stock solution of 0.35 M $CrCl_2(OH_2)_4^+$ has been aquating. The amounts of $CrCl_2(OH_2)_4^+$, $CrCl(OH_2)_5^{2+}$, and $Cr(OH_2)_6^{3+}$ that are now present in solution depend upon the length of time during which the solution has been standing. This should be a few hours. (If allowed to stand overnight, only $Cr(OH_2)_6^{3+}$ will remain.) The object of this portion of the experiment is to determine which complexes are now present in solution.

Prepare a new column of resin. After diluting 10 ml of the 0.35 M $CrCl_2(OH_2)_4^+$ solution with 10 ml of distilled water, pour it on the column. First elute any $CrCl_2(OH_2)_4^+$ that might be present with 0.1 M $HClO_4$. Collect the most intensely colored fraction for identification by its visible spectrum. Next elute $CrCl(OH_2)_5^{2+}$ with 1.0 M $HClO_4$ and collect a fraction for spectral identification. Finally, 3.0 M $HClO_4$ is used to elute any $Cr(OH_2)_6^{3+}$ that may have formed. Record its spectrum.

By comparing the spectra of the fractions obtained in this separation with those determined for the known compounds in the previous section, note which complexes are now present in the solution and roughly which complexes are in greatest and least abundance.

REPORT

Include the following:
1. Spectra of $CrCl_2(OH_2)_4^+$, $CrCl(OH_2)_5^{2+}$, and $Cr(OH_2)_6^{3+}$.
2. By noting the shifts in the positions of the absorptions in these spectra, comment on the relative ligand field strengths of Cl^- and H_2O.
3. The species and their relative concentrations in the aquated 0.35 M $CrCl_2(OH_2)_4^+$ solution.

QUESTIONS

1. The aquation of $CrCl_2(OH_2)_4^+$ to $CrCl(OH_2)_5^{2+}$ is catalyzed by Cr^{2+}. Write a reasonable mechanism that will account for the catalytic role of Cr^{2+}.
2. The conversion of $CrCl(OH_2)_5^{2+}$ to $Cr(OH_2)_6^{3+}$ in water is catalyzed by Hg^{2+}. Write a mechanism for this catalysis.
3. Qualitatively account for the observation that the uncatalyzed aquation of $CrCl_2(OH_2)_4^+$ to $CrCl(OH_2)_5^{2+}$ occurs much faster than that of $CrCl(OH_2)_5^{2+}$ to $Cr(OH_2)_6^{3+}$ under the same conditions.
4. A 10 ml solution of 0.5 M $CrCl(OH_2)_5^{2+}$ is allowed to aquate to $Cr(OH_2)_6^{3+}$. To determine an approximate rate of reaction, the amounts of $CrCl(OH_2)_5^{2+}$ and $Cr(OH_2)_6^{3+}$ present after a certain period of time are evaluated. This is done by pouring the solution on a cation exchange resin in the H^+ form and then titrating the displaced H^+ with base. If 80 ml of 0.15 M NaOH is required to neutralize the liberated H^+, what was the concentration of $CrCl(OH_2)_5^{2+}$ and $Cr(OH_2)_6^{3+}$ in the solution?
5. Explain how you would separate $K_4[Fe(CN)_6]$ from $K_3[Fe(CN)_6]$.
6. Why was perchloric acid rather than HCl used to elute the Cr(III) complexes from the column?
7. Why was visible rather than infrared spectroscopy used to characterize the complexes in this experiment?

INDEPENDENT STUDIES

A. Prepare, isolate, and characterize $[CrCl(OH_2)_5]Cl_2 \cdot H_2O$ and $[Cr(OH_2)_6]Cl_3$. (J. P. Barbier, C. Kappenstein, and R. Hugel, *J. Chem. Educ.*, *49*, 204 (1972).)
B. Determine the rate of aquation of $CrCl(OH_2)_5^{2+}$ to give $Cr(OH_2)_6^{3+}$. (T. W. Swaddle and E. L. King, *Inorg. Chem.*, *4*, 532 (1965).)
C. Prepare and separate by ion exchange chromatography the *cis* and *trans* isomers of $Co(IDA)_2^-$, where $IDA^{2-} = HN(CH_2CO_2^-)_2$. (J. A. Weyh, *J. Chem. Educ.*, *47*, 715 (1970).)

REFERENCES

CrCl$_2$(OH$_2$)$_4$$^+$, CrCl(OH$_2$)$_5$$^{2+}$, and Cr(OH$_2$)$_6$$^{3+}$

I. G. Dance and H. C. Freeman, *Inorg. Chem., 4,* 1555 (1965).

J. H. Espenson and J. P. Birk, *Inorg. Chem., 4,* 527 (1965).

J. H. Espenson and S. G. Slocum, *Inorg. Chem., 6,* 906, (1967).

J. E. Finholt, K. G. Caulton, and W. J. Libbey, *Inorg. Chem., 3,* 1801 (1964).

H. B. Johnson and W. L. Reynolds, *Inorg. Chem., 2,* 468 (1963).

T. W. Swaddle and E. L. King, *Inorg. Chem., 4,* 532 (1965).

Ion Exchange Techniques

E. W. Berg, *Physical and Chemical Methods of Separation,* McGraw-Hill, New York, 1963.

L. F. Druding and G. B. Kauffman, *Coord. Chem. Rev., 3,* 409 (1968). Chromatography of coordination compounds.

R. Kunin, *Ion Exchange Resins,* 2nd Ed., John Wiley and Sons, New York, 1958.

W. Rieman and H. F. Walton, *Ion Exchange in Analytical Chemistry,* Pergamon, New York, 1970.

O. Samuelson, *Ion Exchange Separations in Analytical Chemistry,* John Wiley and Sons, New York, 1963.

For Coordination Chemistry References, See Experiment 1, p. 25, and for Mechanisms of Inorganic Reactions, See Experiment 2, p. 32.

Optical Isomers of Co(en)$_3{}^{3+}$

Optical activity is frequently associated with organic molecules containing an asymmetric carbon atom, as for example in lactic acid,

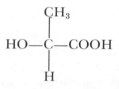

Optical activity, however, is a far more general phenomenon and may be found in any molecule that cannot be superimposed upon its mirror image. In inorganic chemistry, the range of structures that exist as nonsuperimposable mirror images is exceedingly large. While there are many examples of tetrahedral inorganic compounds that have been resolved into their enantiomers, optical activity in octahedral transition metal complexes has been studied far more extensively.

For decades it has been known that certain octahedral complexes of transition metals could be resolved into two enantiomers. Some of the earliest work in this area was done in 1912 by Alfred Werner on Co(en)$_3{}^{3+}$ (where en = $NH_2CH_2CH_2NH_2$). The enantiomers of Co(en)$_3{}^{3+}$ that he resolved were assumed to have the structures shown in Figure 8–1. One of the isomers rotates plane polarized light toward the right (dextrorotatory) while the other isomer rotates the light by the same amount in the opposite (levorotatory) direction. These directions are designated (+) and (−) (or sometimes d and l), respectively. Because of the availability of sodium as a light source, light of 589.3 nm wavelength (the sodium D line) is frequently used in the determination of the rotations. Passing this light through a polarizing prism gives plane polarized light, whose electric field variation is shown in Figure 8–2.

Optically active materials have the ability to rotate the plane of light to the right or left to a greater or lesser angle depending on the nature of the substance. In order for the rotated polarized light to pass

71

Optical Isomers of Co(en)$_3^{3+}$

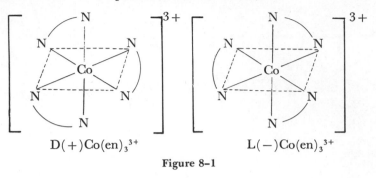

$$D(+)Co(en)_3^{3+} \qquad\qquad L(-)Co(en)_3^{3+}$$

Figure 8–1

through the analyzer prism, this prism must be rotated, relative to the polarizing prism, to the right or left by an angle that is equal to the rotation caused by the sample. Thus, the direction and number of degrees of rotation may be measured experimentally. As in any form of spectroscopy, the size of the rotation depends not only on the nature of the optically active material but also on the length, l, of the light path through the sample and the concentration, c, of the sample in a solvent. To standardize the units for expressing rotations, the specific rotation $[\alpha]_\lambda$ has been defined as the rotation produced by a solution containing 1 g of solute per ml of solution and having a light path length of 1 decimeter.

$$[\alpha]_\lambda = \frac{\alpha}{lc} \tag{1}$$

The wavelength, λ, of light is also specified. Using the sodium D line, the specific rotation is designated $[\alpha]_D$ or $[\alpha]_{589.3}$. In equation (1), α is given in degrees, l in decimeters, and c in grams per ml of solution. A unit that is frequently of more value for comparison between compounds is the molecular rotation, $[M]_\lambda$.

$$[M]_\lambda = \frac{M[\alpha]_\lambda}{100}$$

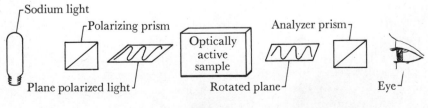

Figure 8–2

Since M is the molecular weight of the substance, $[M]_\lambda$ is a relative measure of its rotatory power on a molecular basis.

The rotatory power of a substance varies with the wavelength of the light employed. Thus, it is usually observed that while a molecule may be dextrorotatory toward light of 589.3 nm wavelength, it is levorotatory at other wavelengths. The values of $[M]_\lambda$ as a function of wavelength are shown for (+)Co(en)$_3$$^{3+}$ in Figure 8–3. This plot is called an optical rotatory dispersion (ORD) curve.

Since the values of $[M]_\lambda$ for enantiomers at any given wavelength are the same but of opposite sign, the ORD curve of (−)Co(en)$_3$$^{3+}$ may be obtained by rotating the curve for (+)Co(en)$_3$$^{3+}$ by 180° around the 0° line in the figure.

Although its optical activity indicated that (+)Co(en)$_3$$^{3+}$ must have one of the structures shown in Figure 8–1, the correct structure was not determined until 1954. By a special x-ray technique a Japanese research group established the absolute configuration of (+)Co(en)$_3$$^{3+}$ as being that shown on the left in Figure 8–1. To show that the absolute configuration is known, the convention of using D to designate this configuration has generally been adopted. Its mirror image, (−)Co(en)$_3$$^{3+}$, must then have the absolute configuration shown on the right of Figure 8–1. It is labeled L(−)Co(en)$_3$$^{3+}$.

While there must be a relationship between the optical activity of an enantiomer and its absolute configuration, the theory of optical

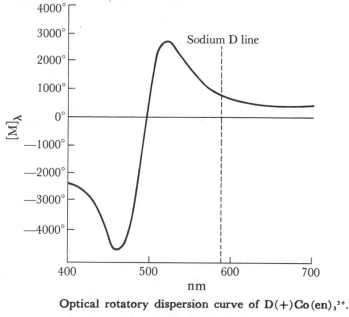

Optical rotatory dispersion curve of D(+)Co(en)$_3$$^{3+}$.

Figure 8–3

activity is sufficiently complex that calculations of absolute configurations from ORD data are highly unreliable. Empirical correlations of ORD curves of very similar molecules have, however, allowed the assignment of absolute configurations to other metal complexes. Such an assignment has been attempted for $Co(en)_2(-)(pn)^{3+}$ (Figure 8-4), where $(-)$pn is levorotatory 1,2-diaminopropane. That the ORD curve (Figure 8-5) of $(+)Co(en)_2(-)(pn)^{3+}$ is virtually the same as that of $D(+)Co(en)_3^{3+}$ suggests that the arrangement of the ligands is the same in $(+)Co(en)_2$-$(-)(pn)^{3+}$ as in $D(+)Co(en)_3^{3+}$. Hence, its absolute configuration is presumed to be that on the left in Figure 8-4.

The use of ORD to establish absolute configurations becomes less reliable as the difference between the electronic state of the standard $D(+)Co(en)_3^{3+}$ and the unknown becomes greater. Thus, while the basic shape of the ORD curve of $(+)cis$-$Co(en)_2(NH_3)_2^{3+}$ is very similar to that of $D(+)Co(en)_3^{3+}$, the values of $[M]_\lambda$ are in general much smaller. The discrepancy in curves becomes even larger for $(+)cis$-$Co(en)_2Cl_2^+$. The question arises as to how similar the curves must be in order to indicate the same absolute configuration. Too few absolute configurations have been established by x-ray methods to permit one to answer this question. Of the vast number of metal complexes that have been resolved into enantiomers, relatively few absolute configurations have been determined, but there is little doubt that this will be an area of active research in the future.

The preparation, resolution, and characterization of the optical isomers of $Co(en)_3^{3+}$ are the objects of this experiment. The preparation of the complex is very similar to that used in the preparation of $Co(NH_3)_5Cl^{2+}$ in Experiment 1. A solution of Co(II) is oxidized by air in

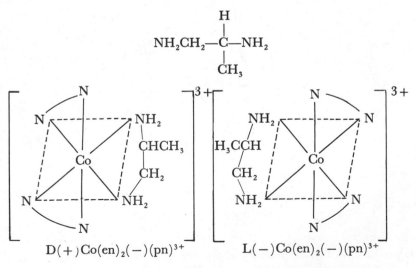

Figure 8-4

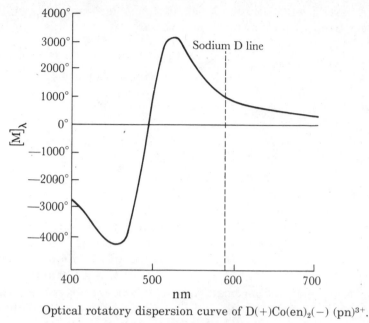

Optical rotatory dispersion curve of D(+)Co(en)$_2$(−) (pn)$^{3+}$.

Figure 8–5

the presence of ethylenediamine, en, and activated charcoal. The activated charcoal catalyzes, by an unknown mechanism, the oxidation of the rapidly formed Co(II) complex, Co(en)$_3$$^{2+}$, to Co(en)$_3$$^{3+}$.

$$CoSO_4 + 3 \text{ en} \rightarrow \left[Co(en)_3\right]SO_4$$

$$4\left[Co(en)_3\right]SO_4 + O_2 + 4HCl \rightarrow 4\left[Co(en)_3\right]SO_4Cl + 2H_2O$$

The resulting $\left[Co(en)_3\right]SO_4Cl$ is not isolated from solution but is immediately resolved by diastereomer-formation with optically active d-tartrate, (+)-tart,

$$
\begin{array}{c}
CO_2^- \\
| \\
HCOH \\
| \\
HOCH \\
| \\
CO_2^-
\end{array}
$$

Diastereomers have differing solubility properties, and with a proper choice of resolving agent it is possible to fractionally crystallize one diastereomer, leaving the other in solution. In this case [(+)Co-

(en)$_3$][(+)-tart]Cl is the least soluble diastereomer and preferentially crystallizes from solution as the pentahydrate.

$$\left.\begin{array}{l}(+)\text{Co(en)}_3{}^{3+}\\(-)\text{Co(en)}_3{}^{3+}\end{array}\right\} +(+)-\text{tart} \xrightarrow[\text{H}_2\text{O}]{\text{Cl}^-} \begin{array}{l}[(+)\text{Co(en)}_3][(+)\text{-tart}]\text{Cl}\cdot 5\text{H}_2\text{O}\downarrow\\[(-)\text{Co(en)}_3][(+)\text{-tart}]\text{Cl}\end{array}$$

The $[(+)\text{Co(en)}_3][(+)\text{-tart}]$Cl is converted to $[(+)\text{Co(en)}_3]\text{I}_3\cdot\text{H}_2\text{O}$ by reaction with I$^-$. The $[\alpha]_\text{D}$ of the product is +89°.

$$[(+)\text{Co(en)}_3][(+)\text{-tart}]\text{Cl} + 3\text{I}^- \rightarrow$$

$$[(+)\text{Co(en)}_3]\text{I}_3\cdot\text{H}_2\text{O}\downarrow + (+)\text{-tart} + \text{Cl}^-$$

The other optical isomer, $[(-)\text{Co(en)}_3]\text{I}_3$, is obtained by adding I$^-$ to the solution from which $[(+)\text{Co(en)}_3][(+)\text{-tart}]\text{Cl}\cdot 5\text{H}_2\text{O}$ was previously precipitated. The solid that precipitates with I$^-$ is a mixture of crystals of the racemate, (+) and $(-)[\text{Co(en)}_3]\text{I}_3\cdot\text{H}_2\text{O}$, and of crystals of pure $[(-)\text{Co(en)}_3]\text{I}_3\cdot\text{H}_2\text{O}$. The $[(-)\text{Co(en)}_3]\text{I}_3\cdot\text{H}_2\text{O}$ is much more soluble in warm water than the racemate and may be extracted into solution, which on cooling reprecipitates the desired enantiomer, $[(-)\text{Co(en)}_3]\text{I}_3\cdot\text{H}_2\text{O}$, whose $[\alpha]_\text{D} = -89°$. The optical purities of the isolated (+) and (−) enantiomers are to be evaluated by measuring their specific rotations.

Finally, it will be shown that the resolved compound may be racemized by boiling an aqueous solution of one of the enantiomers in the presence of activated charcoal.

EXPERIMENTAL PROCEDURE

Preparation of the Resolving Agent, Barium d-Tartrate

Prepare solutions of BaCl$_2$ and of d-tartaric acid by dissolving 12.2 g (50 mmoles) of BaCl$_2\cdot$2H$_2$O in a minimum amount of water and 7.5 g (50 mmoles) of d-tartaric acid in water. After heating these solutions to 90°C, mix them and add the base ethylenediamine until the solution is neutral. Allow the solution to cool to room temperature. Filter the precipitate and wash with warm water.

Preparation and Resolution of Co(en)$_3{}^{3+}$

Prepare in a filter flask a solution containing 10.3 g (170 mmoles, 11.5 ml) of ethylenediamine (en) in 25 ml of water. After cooling the solution in an ice bath, add 5 ml of concentrated (12 M) HCl, 14 g (50 mmoles) of CoSO$_4\cdot$7H$_2$O dissolved in 25 ml of cold water, and 2 g of

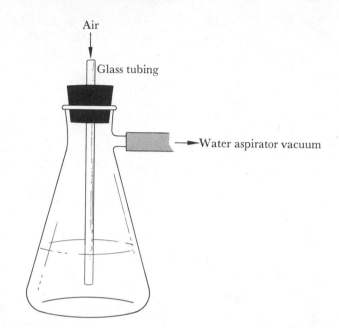

Figure 8–6

activated charcoal. Bubble a rapid stream of air through this solution for 4 hours by pulling a water-aspirator vacuum on the filter flask as shown in Figure 8–6.

Then add dilute HCl or ethylenediamine, as required, to the solution until its pH is 7.0–7.5. Heat the mixture in an evaporating dish on a steam bath for 15 minutes. Cool the solution to room temperature, filter off the charcoal, and wash with 10 ml of water on the filter. Add the wash to the filtrate.

To the Co(en)$_3$$^{3+}$ solution just prepared, add all of the barium d-tartrate prepared previously. After heating the mixture on a steam bath for 30 minutes with vigorous stirring, filter off the precipitated BaSO$_4$ and wash with a small amount of hot water. Evaporate the solution on a hot plate or with a burner to 50 ml, and allow the crystals of $[(+)\text{Co-}$(en)$_3][(+)\text{-tart}]\text{Cl}\cdot5\text{H}_2\text{O}$ to precipitate overnight. Filter off the crystals and save the filtrate for later isolation of the $(-)$ Co(en)$_3$$^{3+}$ enantiomer. Wash the crystals with a 40 per cent (by volume) ethanol-water solution, and recrystallize the product by dissolving it in 15 ml of hot water followed by cooling in ice. Wash the crystals with 40 per cent ethanol-water and then with absolute ethanol. Air-dry and determine the yield of $[(+)\text{Co(en)}_3][(+)\text{-tart}]\text{Cl}\cdot5\text{H}_2\text{O}$.

To determine the specific rotation of this compound, dilute approximately 0.5 g of the sample to a solution volume of 10 ml. Introduce this solution into a 1 decimeter polarimeter tube, tilting the tube as necessary to remove bubbles from the light path. Following the manufacturer's instructions, first adjust the polarimeter so that the angle of

rotation is zero when there is no sample in the instrument. Then introduce the polarimeter tube containing the sample solution into the polarimeter, and measure the sign and magnitude of the rotation. If the
instrument requires visual matching of fields, darkening the room is
helpful. Calculate $[\alpha]_D$ for $[(+)Co(en)_3][(+)\text{-tart}]Cl \cdot 5H_2O$.

To convert the diastereomer to $[(+)Co(en)_3]I_3 \cdot H_2O$, dissolve the
$[(+)Co(en)_3][(+)\text{-tart}]Cl \cdot 5H_2O$ in 15 ml of hot water and add 0.25 ml
of concentrated ammonia (15 M) solution. With stirring, add a solution
of 17 g (113 mmoles) of NaI dissolved in 7 ml of hot water. After cooling
in an ice bath, suction filter and wash the crystals with an ice cold solution of 3 g of NaI in 10 ml of water to remove the tartrate. After washing
with ethanol and finally with acetone, allow the $[(+)Co(en)_3]I_3 \cdot H_2O$
to air-dry and determine the yield. Measure its $[\alpha]_D$ using a solution of
approximately 0.5 g of sample in 10 ml of water.

To isolate $[(-)Co(en)_3]I_3 \cdot H_2O$, add 0.25 ml of concentrated NH_3
solution to the filtrate from which $[(+)Co(en)_3][(+)\text{-tart}]Cl \cdot 5H_2O$
was precipitated (see previous discussion). Heat the solution to 80°C
and add with stirring 17 g (113 mmoles) of NaI. Upon cooling in an
ice bath, impure $[(-)Co(en)_3]I_3 \cdot H_2O$ precipitates and is filtered and
washed with a solution of 3 g of NaI dissolved in 10 ml of water. To
purify, dissolve the precipitate, with stirring, in 35 ml of water at 50°C.
Filter off the undissolved racemate and add 5 g of NaI to the filtrate.
Crystallization of $[(-)Co(en)_3]I_3 \cdot H_2O$ occurs on cooling. Filter the precipitate, wash with ethanol and then with acetone, and finally air-dry.
Determine the yield and evaluate $[\alpha]_D$.

Racemization of $(+)Co(en)_3{}^{3+}$ or $(-)Co(en)_3{}^{3+}$

Dissolve approximately 1 g of either $[(+)Co(en)_3]I_3 \cdot H_2O$ *or*
$[(-)Co(en)_3]I_3 \cdot H_2O$ in a minimum volume of warm water. Add a small
amount of activated charcoal and boil the solution for approximately 30
minutes. Then filter the solution while hot, and add a few grams of NaI to
aid in the precipitation of the racemate. Wash with alcohol and acetone
and air-dry. Determine $[\alpha]_D$ of the racemized $[Co(en)_3]I_3 \cdot H_2O$.

REPORT

Include the following:
1. Percentage yields of $[(+)Co(en)_3][(+)\text{-tart}]Cl \cdot 5H_2O$, $[(+)Co(en)_3]\text{-}$
   $I_3 \cdot H_2O$, and $[(-)Co(en)_3]I_3 \cdot H_2O$.
2. $[\alpha]_D$ and $[M]_D$ values for the above complexes. (If an ORD spectrometer is available, record their ORD curves.)
3. If $[\alpha]_D$ of pure $[(+)Co(en)_3]I_3 \cdot H_2O$ is +89°, what percentage of
   your sample of this compound is actually this enantiomer? Assume

that the only impurity is $[(-)Co(en)_3]I_3 \cdot H_2O$. Do the same calculation for your sample of $[(-)Co(en)_3]I_3 \cdot H_2O$.

4. $[\alpha]_D$ of the $[Co(en)_3]I_3 \cdot H_2O$, which is isolated from boiling a solution of (+) or (−) Co(en)$_3$$^{3+}$ with activated charcoal.

QUESTIONS

1. Plot an ORD curve for (−)Co(en)$_3$$^{3+}$ analogous to that given in Figure 8–3 for (+)Co(en)$_3$$^{3+}$.
2. If you were to resolve an unknown complex, M(en)$_3$$^{3+}$, how would you know whether or not your resolved products were optically pure?
3. Draw structures of the geometrical and optical isomers of Co(gly)$_3$, where gly = $NH_2CH_2COO^-$.
4. Why is it not possible to resolve Co(en)$_3$$^{2+}$?
5. Draw structures of the optical isomers of Co(EDTA)$^-$, where EDTA = $(^-O_2CCH_2)_2NCH_2CH_2N(CH_2CO_2{}^-)_2$.
6. In the preparation of barium d-tartrate, what was the purpose of adding ethylenediamine?
7. In the purification of both (+) and (−) $[Co(en)_3]I_3 \cdot H_2O$, the compounds were washed with water containing NaI. What was the purpose of the NaI?
8. Outline methods for analyzing $[(+)Co(en)_3]I_3 \cdot H_2O$ for its percentage Co and I content.

INDEPENDENT STUDIES

A. Analyze $[Co(en)_3]I_3 \cdot H_2O$ for its percentage Co and/or I content.
B. Confirm the 4-ion nature of $[Co(en)_3]I_3 \cdot H_2O$ by measuring its molar conductance.
C. Measure the proton nmr spectrum of $[Co(en)_3]I_3 \cdot H_2O$ in D$_2$O solvent. (J. K. Beattie, *Accounts Chem. Res., 4*, 253 (1971).)
D. Prepare and resolve $[Ni(o\text{-phen})_3](ClO_4)_2$, where o-phen is 1,10-phenanthroline, into its d and l enantiomers. (G. B. Kauffman and L. T. Takahashi, *Inorganic Syntheses, Vol. 8*, McGraw-Hill, New York, 1966, p. 227.)
E. Record and compare the ultraviolet-visible spectra of the optical isomers of $[Co(en)_3]I_3 \cdot H_2O$.

REFERENCES

Resolution of Co(en)$_3$$^{3+}$

J. A. Broomhead, F. P. Dwyer, and J. W. Hogarth, *Inorganic Syntheses, Vol. 6*, McGraw-Hill, New York, 1960, p. 183.
C. F. Bell, *Syntheses and Physical Studies of Inorganic Compounds*, Pergamon Press, New York, 1972, Chap. 22. A summary of physical studies of Co(en)$_3$$^{3+}$.

Techniques of Optical Activity

F. Woldbye in *Technique of Inorganic Chemistry, Vol. IV,* H. B. Jonassen and A. Weissberger, eds., Interscience Publishers, New York, 1965, p. 249. Instrumental methods, theory, and survey of the literature of optical rotatory dispersion and circular dichroism.

Optical Activity

J. G. Foss, *J. Chem. Educ., 40,* 592 (1963). Absorption, dispersion, circular dichroism, and rotatory dispersion.
R. D. Gillard in *Progress in Inorganic Chemistry, Vol. 7,* F. A. Cotton, ed., Interscience Publishers, New York, 1966, p. 215. Optical rotatory dispersion and circular dichroism studies of coordination compounds.
C. J. Hawkins, *Absolute Configuration of Metal Complexes,* Wiley-Interscience, New York, 1971. Theory of optical activity in transition metal complexes.
G. B. Kauffman, *Coord. Chem. Rev., 12,* 105 (1974). Historical account of Alfred Werner's research on optically active coordination compounds.
A. M. Sargeson in *Chelating Agents and Metal Chelates,* F. P. Dwyer and D. P. Mellor, eds., Academic Press, New York, 1964, p. 183. A review of many aspects of optically active transition metal complexes.
N. Serpone and D. G. Bickley in *Progress in Inorganic Chemistry, Vol. 17,* J. O. Edwards, ed., Interscience Publishers, New York, 1972, p. 391. Review of rates and mechanisms of racemization in six-coordinate chelate compounds.
Y. Saito, *Coord. Chem. Rev., 13,* 305 (1974). Review of absolute configurations of complexes determined by x-ray techniques.

[PCl₄][SbCl₆] and [P(C₆H₅)₃Cl][SbCl₆]

The group V elements P, As, Sb, and Bi form binary halides of the composition MX_3 with all four halides, F, Cl, Br, and I. In the gas phase and usually in the solid state these molecules have a pyramidal structure, as shown for PCl_3:

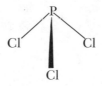

Like the halides of many of the group IV elements, e.g., $SnCl_4$ (see Experiment 17), these halides react with Grignard reagents to form the organocompounds MR_3. While its synthesis will not be carried out in this experiment, $P(C_6H_5)_3$ is readily prepared by the reaction:

$$PCl_3 + 3C_6H_5MgBr \rightarrow P(C_6H_5)_3 + 3MgBrCl$$

Triphenylphosphine, $P(C_6H_5)_3$, is commercially available and will be used as a starting material in one of the reactions in this experiment.

The halides PCl_3 and $SbCl_3$ react with Cl_2 to form the halides PCl_5 and $SbCl_5$ in which P and Sb are present in the $+5$ oxidation state. (Strangely, the analogous $AsCl_5$ is still unknown.)

$$PCl_3 + Cl_2 \rightleftharpoons PCl_5$$

$$SbCl_3 + Cl_2 \rightleftharpoons SbCl_5$$

In the gas phase, PCl_5 and $SbCl_5$ dissociate appreciably to the trihalide and Cl_2, as indicated by the reverse arrows in the above equations. At room temperature PCl_5 is a solid and $SbCl_5$ is a liquid. Both are very

81

hygroscopic; PCl_5, for example, reacts with a small amount of water to yield the tetrahedral molecule, $OPCl_3$,

$$PCl_5 + H_2O \rightarrow OPCl_3 + 2HCl$$

In an excess of water, the remaining Cl groups are replaced by -OH to give phosphoric acid:

$$OPCl_3 + 3H_2O \rightarrow OP(OH)_3 + 3HCl$$

The extreme reactivity of PCl_5 and $SbCl_5$ with moisture requires that studies of these compounds be conducted in a dry box or dry bag. Because they are inexpensive and convenient to use, polyethylene dry bags will be utilized in this experiment.

It has been established by x-ray and electron diffraction as well as Raman spectroscopic studies that $SbCl_5$ has a trigonal bipyramidal structure,

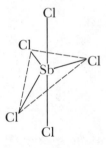

in all three states—solid, liquid, and gas. Phosphorus pentachloride, on the other hand, has the trigonal bipyramidal structure in the gas phase but exists as the ionic $[PCl_4^+][PCl_6^-]$ in the solid state. Discrete tetrahedral PCl_4^+ cations and octahedral PCl_6^- anions are packed into the crystalline lattice. To further complicate the structural features of PCl_5, there is evidence that in CCl_4 solution it exists as the dimer, $(PCl_5)_2$, which presumably has the structure

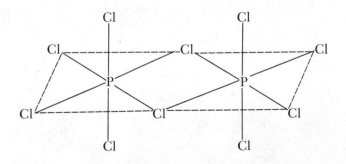

with two bridging Cl atoms. The tendency of PCl$_5$ to ionize may also be illustrated by its reaction with the strong Cl$^-$ acceptor AlCl$_3$:

$$PCl_5 + AlCl_3 \rightarrow [PCl_4][AlCl_4]$$

While the existence of PCl$_4^+$ has been established in several compounds, the PCl$_6^-$ anion is known in only a few cases. The instability of PCl$_6^-$ contrasts with the familiar PF$_6^-$ ion, which is stable in water and air.

The increased size of the Sb atom in SbCl$_5$ is apparently sufficient to allow the addition of a chloride ion to give the well-known octahedral SbCl$_6^-$ ion:

$$SbCl_5 + Cl^- \rightarrow SbCl_6^-$$

Moreover, SbCl$_5$ is such a powerful Cl$^-$ acceptor that it abstracts Cl$^-$ even from acyl halides, e.g.,

$$SbCl_5 + CH_3-C\overset{\displaystyle O}{\underset{\displaystyle Cl}{\big<}} \rightarrow \left[CH_3-C\equiv O^+\right]\left[SbCl_6^-\right]$$

In general, SbCl$_5$ is a very strong Lewis acid and reacts with numerous Lewis bases, L, to form the octahedral adducts, SbCl$_5$L.

$$SbCl_5 + L \rightarrow SbCl_5L$$

$$L = (C_2H_5)_2O, \text{ pyridine}, (CH_3)_2SO, \text{ or } Cl_3PO$$

In this experiment, the strong Lewis acid properties of SbCl$_5$ will be demonstrated by its reaction with PCl$_5$ to give [PCl$_4^+$][SbCl$_6^-$]:

$$PCl_5 + SbCl_5 \rightarrow \left[PCl_4^+\right]\left[SbCl_6^-\right]$$

In principle, there is a possibility that the product could be $\left[SbCl_4^+\right]$ $\left[PCl_6^-\right]$, since there is evidence that the SbCl$_4^+$ ion exists in certain other compounds. The relatively high stabilities of the PCl$_4^+$ and SbCl$_6^-$ ions, however, are apparently sufficient to cause the transfer of Cl$^-$ from PCl$_5$ to SbCl$_5$ rather than from SbCl$_5$ to PCl$_5$.

Not only is SbCl$_5$ a strong Lewis acid, but it may also act as a powerful chlorinating agent. Both of these properties are involved in the reaction of SbCl$_5$ with P(C$_6$H$_5$)$_3$; this reaction, which will be carried out in this experiment, gives the salt $\left[(C_6H_5)_3PCl^+\right]\left[SbCl_6^-\right]$, m.p. 172°C,

$$(C_6H_5)_3P + 2SbCl_5 \rightarrow \left[(C_6H_5)_3PCl^+\right]\left[SbCl_6^-\right] + SbCl_3$$

This net reaction probably proceeds in two steps; first the chlorination of $P(C_6H_5)_3$,

$$(C_6H_5)_3P + SbCl_5 \rightarrow (C_6H_5)_3PCl_2 + SbCl_3,$$

and second the Cl^- transfer to $SbCl_5$,

$$(C_6H_5)_3PCl_2 + SbCl_5 \rightarrow [(C_6H_5)_3PCl^+][SbCl_6^-].$$

While the intermediate $(C_6H_5)_3PCl_2$ cannot be isolated from this reaction, it may be prepared from $(C_6H_5)_3P$ and Cl_2. Infrared evidence suggests that it exists in an ionic form, $[(C_6H_5)_3PCl^+]Cl^-$. The $(C_6H_5)_3PCl_2$ prepared from $(C_6H_5)_3P$ and Cl_2 reacts with $SbCl_5$, to form $[(C_6H_5)_3PCl^+][SbCl_6^-]$, thus supporting the preceding two-step sequence. In terms of these two steps, the overall reaction is easily rationalized.

The ionic composition of $[(C_6H_5)_3PCl][SbCl_6]$ will be established by determining its molar conductance, Λ_M, in nitrobenzene solvent by the procedure given in Experiment 1, p. 17. Previously determined molar conductances of 2- and 3-ion conductors at 25° are

Number of Ions	Λ_M
2	20–30
3	50–60

(See Appendix 2 for full listing of molar conductances in solvents.) Measurement of the molar conductance of the exceedingly hygroscopic $[PCl_4][SbCl_6]$ would require special precautions to exclude moisture. Reaction with even a small amount of water would produce ions and render the molar conductance meaningless.

EXPERIMENTAL PROCEDURE

Both PCl_5 and $SbCl_5$ are very sensitive to moisture. Depending upon the atmospheric humidity, their handling will require more or less rigorous protection from air. Their reactions will be conducted in a dry bag. Triphenylphosphine, $(C_6H_5)_3P$, is a stable solid. Although both products are very hygroscopic, $[PCl_4][SbCl_6]$ is much more sensitive to moisture than $[(C_6H_5)_3PCl][SbCl_6]$. The use of a dry bag requires planning of experimental operations before they are actually carried out in the bag. To stabilize the bag, it is convenient to put a couple of ring stand bases in the bag before introducing your equipment.

Connect the polyethylene dry bag (or glove bag) (Figure 9–1) to a nitrogen cylinder and allow the dry gas to flush the bag for about 5 minutes; then place all chemicals and equipment needed for the prep-

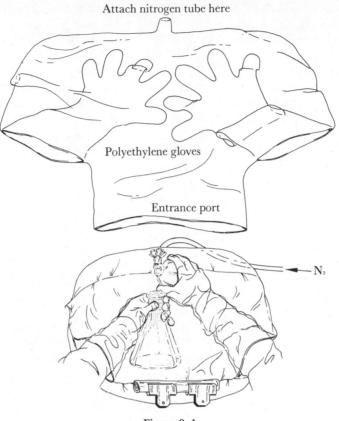

Attach nitrogen tube here

Polyethylene gloves

Entrance port

N$_2$

Figure 9–1

aration of both $\left[PCl_4\right]\left[SbCl_6\right]$ and $\left[(C_6H_5)_3PCl\right]\left[SbCl_6\right]$ in the bag. This will include:

    1—spatula
    1—10 ml graduated cylinder
    1—50 ml graduated cylinder
    1—small glass stirring rod
    2—medium glass frits
    1—rubber frit holder or adapter
    2—125 ml filter flasks
    1—suction bulb
    2—125 ml Erlenmeyer flasks
    4—rubber stoppers for the filter and Erlenmeyer flasks
    2—drying tubes (Figure 6–4)
    a small reagent bottle of SbCl$_5$
    a small bottle or can of anhydrous (C$_2$H$_5$)$_2$O
    a stoppered 125 ml Erlenmeyer flask containing 100 ml of CH$_2$Cl$_2$

a stoppered 125 ml Erlenmeyer flask containing 10 mmoles (2.6 g) of $(C_6H_5)_3P$

a stoppered 125 ml Erlenmeyer flask containing 10 mmoles (2.1 g) of PCl_5

Since two syntheses will be conducted in the bag, the flasks and drying tubes should be labeled to avoid confusion in the preparations. The last step in putting these materials in the dry bag should be weighing the PCl_5 into the Erlenmeyer and then placing the flask in the bag. With the entrance port of the bag still open and the nitrogen flowing, push on the top of the bag to make it as small as possible and then roll up the polyethylene at the entrance port and clamp it closed. When the nitrogen has filled the bag to a convenient size (it should be flabby, not completely full), turn off the nitrogen. Powder your hands with talcum powder or wear cotton gloves before putting your hands into the gloves in the bag. Take your time in carrying out the preparations, because all operations are more difficult in the bag, and a spilled solution makes a mess and usually ruins the experiment.

[PCl₄][SbCl₆]

To a solution of 10 mmoles (1.3 ml) of $SbCl_5$ in 10 ml of CH_2Cl_2, *slowly* add a solution of 10 mmoles (2.1 g) of PCl_5 in 50 ml of CH_2Cl_2. (At the same time that you make the $SbCl_5$ solution, it is also convenient to prepare the $SbCl_5$ solution for the $[(C_6H_5)_3PCl][SbCl_6]$ synthesis in another Erlenmeyer flask and stopper it. This will minimize the handling of $SbCl_5$ and the chances of spillage.) Precipitation of the product occurs instantly. It is filtered in the suction filtration apparatus shown in Figure 9–2, in which the suction bulb provides the vacuum.

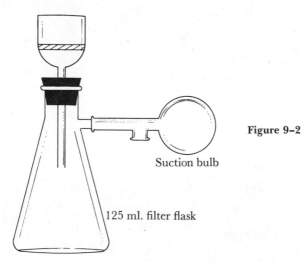

Suction bulb

Figure 9–2

125 ml. filter flask

Wash the solid once with ~10 ml of ether. Transfer the solid to a labeled drying tube (Figure 6–4) and close it. Set it aside and prepare $[(C_6H_5)_3PCl][SbCl_5]$.

[(C₆H₅)₃PCl][SbCl₆]

To a solution of 20 mmoles (2.6 ml) of $SbCl_5$ in 10 ml of CH_2Cl_2, *slowly* add a solution of 10 mmoles (2.6 g) of $(C_6H_5)_3P$ in 20 ml of CH_2Cl_2. (This reaction is exothermic, and the CH_2Cl_2 will boil, b.p. 40°C, if the addition is made too fast.) After the solution has cooled to room temperature, add 75 ml of ethyl ether while swirling the flask. Precipitation of the product will occur as the last of the ether is added. Suction filter and rinse the product with ~10 ml of ether as was done in the previous preparation. Transfer the $[(C_6H_5)_3PCl][SbCl_6]$ to the drying tube and close it.

Remove the drying tubes from the bag. Attach them to the utility vacuum line (Introduction, p. 8) and allow them to dry for 1 hour under a dynamic vacuum. Then quickly transfer the samples to closed preweighed weighing bottles and determine the yield of both [PCl₄]-[SbCl₆] and [(C₆H₅)₃PCl][SbCl₆]. The products should be stored in a desiccator. Measure the melting point of $[(C_6H_5)_3PCl][SbCl_6]$.

Determine the molar conductance (Experiment 1, p. 17) at 25°C of an approximately 3×10^{-3} M nitrobenzene solution of [(C₆H₅)₃PCl]-[SbCl₆] to establish the ionic character of the compound. (Because the [(C₆H₅)₃PCl][SbCl₆] solution will slowly hydrolyze in air, the conductance measurement should be done immediately after the solution is prepared.) The cell constant, k, must be determined using 0.02 M KCl in water. This constant together with the resistance measurement on the nitrobenzene solution is then used to calculate the molar conductance of $[(C_6H_5)_3PCl][SbCl_6]$.

To examine the sensitivity of these compounds toward water, place watch glasses containing PCl₅, SbCl₅, [PCl₄][SbCl₆], and [(C₆H₅)₃PCl]-[SbCl₆] on the bench top. Note what happens and determine which substances react most quickly with atmospheric moisture.

REPORT

Include the following:
1. Yields of both $[PCl_4][SbCl_6]$ and $[(C_6H_5)_3PCl][SbCl_6]$.
2. Melting point of [(C₆H₅)₃PCl][SbCl₆].
3. Value of Λ_M for [(C₆H₅)₃PCl][SbCl₆]. Does it indicate that the compound is a 2-ion conductor?
4. Observations and balanced equations for the probable reactions of PCl₅, SbCl₅, [PCl₄][SbCl₆], and [(C₆H₅)₃PCl][SbCl₆] with atmospheric moisture.

QUESTIONS

1. Outline a method for analyzing $[PCl_4][SbCl_6]$ for its percentage Cl content.
2. Realizing the extreme sensitivity of $[PCl_4][SbCl_6]$ to moisture, what precautions would you take in measuring its molar conductance in nitrobenzene solution?
3. What experiments would you conduct to determine whether PCl_5 was monomeric, dimeric, or ionic in CCl_4 solution?
4. The heavier of two analogous compounds usually has the higher melting point. For example, the melting point of $SbCl_3$ (73°C) is higher than that of PCl_3 ($-112°C$). Account for the fact that the melting point of $SbCl_5$ (2.3°C) is lower than that of PCl_5 (167°C, sublimes).
5. Draw structural formulas for $[PCl_4][SbCl_6]$ and $[(C_6H_5)_3PCl]$-$[SbCl_6]$.
6. When making $[PCl_4][SbCl_6]$, how would you know that the dry bag had not been adequately flushed with N_2?
7. Write equations for the preparation of $[(C_6H_5)PCl_3][SbCl_6]$ from PCl_3 and $SbCl_5$.
8. If you were to purify $[(C_6H_5)_3PCl][SbCl_6]$ by recrystallization, what solvent(s) would you use; and how would you do it?

INDEPENDENT STUDIES

A. Analyze $[PCl_4][SbCl_6]$ and/or $[(C_6H_5)_3PCl][SbCl_6]$ for their percentage Cl content.
B. Prepare and characterize adducts of $SbCl_5$ such as (pyridine)$SbCl_5$ (J. C. Hutton and H. W. Webb, *J. Chem. Soc.*, 1518 (1931)) or (dimethylformamide)$SbCl_5$ (G. Hilgetag and H. Teichmann, *Chem. Ber.*, *96*, 1446 (1963)).
C. Prepare $BiBr_4^-$, $BiBr_5^{2-}$, or $BiBr_6^{3-}$ from $BiBr_3$ and Br^-. (R. D. Whealy and J. F. Osborne, *Inorg. Chim. Acta, 4,* 420 (1970).)
D. Prepare the cyclic phosphonitrilic chloride, $(NPCl_2)_3$, from PCl_5 and NH_4Cl. (J. Emsley and P. B. Udy, *J. Chem. Soc. A*, 768 (1971).)

REFERENCES

$[PCl_4][SbCl_6]$ and $[(C_6H_5)_3PCl][SbCl_6]$

C. M. Harris and R. S. Nyholm, *J. Chem. Soc.*, 4375 (1956). Molar conductances of 2- and 3-ion conductors in nitrobenzene.
J. K. Ruff, *Inorg. Chem.*, 2, 813 (1963). Preparation and physical measurements on $[PR_nCl_{4-n}][SbCl_6]$ salts.

Glove Box Techniques

E. C. Ashby and R. D. Schwartz, *J. Chem. Educ.*, *51*, 65 (1974). A glove box with a recirculating system.

C. J. Barton in *Technique of Inorganic Chemistry, Vol. III,* H. B. Jonassen and A. Weissberger, eds., Interscience Publishers, New York, 1963, p. 259.

J. Casteel and E. S. Amis, *J. Chem. Educ., 48,* 460 (1971). A supported glove bag.

L. F. Druding, *J. Chem. Educ., 47,* A815 (1970).

S. G. Shore, *J. Chem. Educ., 39,* 465 (1962). A supported glove bag.

D. F. Shriver, *The Manipulation of Air Sensitive Compounds,* McGraw-Hill, New York, 1969, Chap. 8. Various glove boxes and bags and their use.

Halides of P, As, Sb, and Bi

C. F. Bell, *Syntheses and Physical Studies of Inorganic Compounds,* Pergamon Press, New York, 1972, Chap. 8. A summary of physical studies of PCl_5.

J. W. George in *Progress in Inorganic Chemistry, Vol. II,* F. A. Cotton, ed., Interscience Publishers, New York, 1960, p. 33. Halides and oxyhalides of group V and VI elements.

R. R. Holmes, *J. Chem. Educ., 40,* 125 (1963). Structures and properties of halides of P, As, Sb, and Bi.

L. Kolditz in *Advances in Inorganic Chemistry and Radiochemistry, Vol. 7,* H. J. Emeleus and A. G. Sharpe, eds., Academic Press, New York, 1965, p. 1. Preparations, structures and properties.

L. Kolditz in *Halogen Chemistry, Vol. 2,* V. Gutmann, ed., Academic Press, New York, 1967, p. 115. Halides of As and Sb.

M. Webster, *Chem. Rev., 66,* 87 (1966). Lewis acid properties of group V pentahalides.

N_2O_4

Note: A continuous period of 4 to 5 hours is generally required to complete the preparation of N_2O_4.

Nitrogen forms a variety of oxygen compounds. Among those that are well characterized are N_2O, NO, N_2O_3, NO_2, N_2O_4, and N_2O_5. Depending upon the conditions of temperature, pressure, and concentration of O_2, two or more of these oxides are found to exist in equilibrium. For example, NO reacts readily with O_2 to give NO_2; it also disproportionates spontaneously to yield N_2O and NO_2; and of course, NO_2 is known to dimerize to N_2O_4. While such facile conversions from one oxide to another are frequently inconvenient, they do allow the ready synthesis of one oxide from another.

This experiment requires first the synthesis of N_2O_5 and then its conversion to N_2O_4. Dinitrogen pentoxide may be prepared by dehydrating its corresponding acid, HNO_3:

$$2 \ HNO_3 \rightarrow N_2O_5 + H_2O$$

The strong affinity of N_2O_5 for H_2O requires the use of a very strong dehydrating agent such as phosphorus pentoxide, P_4O_{10}, which is sufficiently deliquescent to remove H_2O from HNO_3:

$$4 \ HNO_3 + P_4O_{10} \rightarrow 4 \ HPO_3 + 2 \ N_2O_5$$

The resulting N_2O_5 is a white solid that sublimes at room temperature. Its structure in the gas phase is

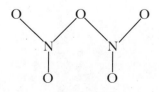

In the solid state, N_2O_5 no longer has a molecular structure but is ionic, and is more accurately formulated as $NO_2^+ \; NO_3^-$. Dimensions of the ions are as follows:

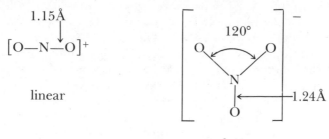

linear

planar

Both the nitronium ion, NO_2^+, and the nitrate ion, NO_3^-, are well known in other types of salts. In the gas phase N_2O_5 is thermodynamically unstable and loses O_2 to form the equilibrium mixture of NO_2 and N_2O_4.

$$2N_2O_5 \rightarrow 2N_2O_4 + O_2$$

While this decomposition occurs slowly even at room temperature, elevated temperatures are more convenient for synthetic purposes. In the present experiment this reaction will be conducted at 260°C in a tube furnace.

The product, N_2O_4, has the planar structure

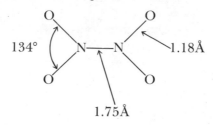

in which the N—N bond is exceedingly long and therefore presumably weak. This indeed seems to be evidenced by its tendency to dissociate into two molecules of brown, paramagnetic NO_2:

$$N_2O_4 \rightleftharpoons 2NO_2$$
colorless   brown

Nitrogen dioxide, NO_2, has a bent structure that is virtually identical

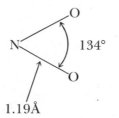

in geometry to half of the N_2O_4 molecule. In the solid state, the NO_2–N_2O_4 equilibrium mixture is all N_2O_4, as indicated by its white color. Liquid N_2O_4, even at its boiling point (21.3°C), contains only 0.1 per cent NO_2. In the gas phase, however, dissociation becomes appreciable. At 27°C, for example, 20 per cent of the N_2O_4 dissociates to NO_2, and at 100°C, 90 per cent exists as NO_2. The increasing concentration of NO_2 in these mixtures is obvious from the increasing intensity of the brown color as the temperature of the gas is raised. The presence of both NO_2 and N_2O_4 will be established in this experiment by examining the infrared spectrum of the gas at room temperature.

Liquid N_2O_4 boils at 21.3°C and freezes at −12.3°C. Although it could be conveniently used as a solvent, its low dielectric constant (2.4 at 18°C) gives it solubilizing properties about the same as those of diethyl ether. In general, ionic reactants are not sufficiently soluble to make N_2O_4 a useful solvent for many reactions. For a few specialized cases, it has been used, however.

There is evidence that N_2O_4, like H_2O, undergoes an autoionization reaction:

$$2H_2O \rightleftharpoons H_3O^+ + OH^-$$

$$N_2O_4 \rightleftharpoons NO^+ + NO_3^-$$

Conductivity studies of liquid N_2O_4 indicate that ionization occurs, however, to an even lesser extent than is observed for H_2O. The presence of NO^+ and NO_3^- in N_2O_4 allows one to account more easily for the reactions of the liquid. For example,

$$KCl + N_2O_4 \rightarrow ClNO + KNO_3$$
$$(NO^+NO_3^-)$$

If N_2O_4 is viewed as $NO^+NO_3^-$, the above reaction might be considered a simple metathesis reaction. Also, its reaction with metallic sodium

$$Na + N_2O_4 \rightarrow NO + NaNO_3$$
$$(NO^+NO_3^-)$$

might be regarded as a reduction of NO^+ by Na. This latter reaction is very similar to that of Cu, as carried out in Experiment 11.

$$Cu + 2N_2O_4 \rightarrow 2NO + Cu(NO_3)_2$$
$$(NO^+NO_3^-)$$

While it is convenient to view N_2O_4 as reacting *via* the ionic $NO^+NO_3^-$

form, this has not been established, and there is even evidence to suggest that the non-ionic isomer, ONONO$_2$, is an intermediate in some reactions.

EXPERIMENTAL PROCEDURE

NO$_2$–N$_2$O$_4$ is exceedingly toxic and is at least as dangerous as carbon monoxide, CO, or hydrogen cyanide, HCN. *All operations involving N$_2$O$_4$ must be conducted in an efficient hood.*

Dinitrogen Tetroxide, N$_2$O$_4$

Because N$_2$O$_4$ attacks rubber, all connections and containers must be glass. The preparation will be carried out in the apparatus shown in Figure 10–1. (Ground glass joints are all 24/40 standard taper.) Make certain that all of the glassware is dry. Fill the drying tube with anhydrous CaSO$_4$. Plug one end of the P$_4$O$_{10}$ drying column with glass wool (handle glass wool with gloves), and *loosely* pack the tube with P$_4$O$_{10}$. After putting 100 g of P$_4$O$_{10}$ in the 3-neck round-bottom flask, assemble the equipment as indicated in Figure 10–1, making certain to lubricate all ground glass joints with silicone grease. Lubricate the ground glass bearing of the stirring shaft with glycerin. Turn on the tube furnace (the furnace used in Experiment 3, p. 36, may be used here) and adjust its temperature to 260°C. (Do not go above this temperature because the

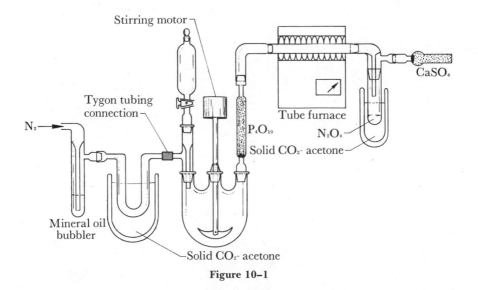

Figure 10–1

decomposition, $2 NO_2 \rightarrow 2 NO + O_2$, becomes appreciable.) At the same time, turn on the dry nitrogen flush and add 60 ml of fuming nitric acid (concentrated HNO_3 will not work) to the addition funnel. Place an acetone bath containing solid CO_2 ($-78°C$) around the U-tube trap. (The purpose of the U-tube trap is to remove any water that the "dry" nitrogen might contain.) Start the stirrer and add the fuming HNO_3 dropwise to the P_4O_{10} at a sufficiently slow rate to prevent distillation of the acid into the connecting tubes. When the brown NO_2—N_2O_4 gas reaches the product trap, immerse the lower half of the trap in an acetone-solid CO_2 Dewar flask. Adjust the N_2 flow to facilitate the condensation of solid, colorless N_2O_4 into the trap.

During the addition of fuming HNO_3, the mixture in the 3-neck flask will become very sticky, and stirring will have to be temporarily suspended. After 25 or 30 ml of the HNO_3 has been added, it becomes more fluid and stirring should be resumed. Continue the dropwise addition of fuming HNO_3 until approximately 55 ml has been added (2 to 3 hours). Allow the N_2 flow to continue until all of the N_2O_4 has been collected. Then remove the product trap and quickly close it with a ground glass stopper. Store it in a freezer or an acetone-solid CO_2 bath. The apparatus in Figure 10–1 should be disassembled immediately after the synthesis is completed; otherwise the ground glass joints that come in contact with the P_4O_{10} or HPO_3 have a notable tendency to freeze. Estimate the volume of N_2O_4 collected and from its density calculate the approximate percentage yield.

The N_2O_4 obtained is sufficiently pure for its infrared spectrum to be examined. This will be done in the gas phase by transferring the N_2O_4 trap to the utility vacuum line (see Introduction), shown in Figure 10–2. Keep it cooled in an acetone-solid CO_2 bath ($-78°C$).

With the N_2O_4 frozen, open stopcocks B, C, and D (close E) and then slowly open A to evacuate the entire system. Note the pressure on

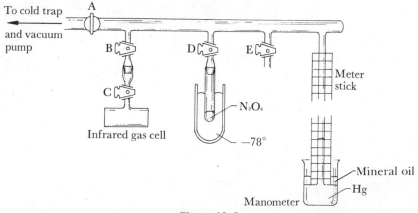

Figure 10–2

the manometer. When the line has been evacuated, close A and remove the acetone-solid CO$_2$ Dewar flask until the Hg manometer has dropped by about 2 cm. Then close D and replace the $-78°$C Dewar flask around the N$_2$O$_4$. An approximate pressure of 2 cm of NO$_2$–N$_2$O$_4$ is required to give a satisfactory infrared spectrum in a 10 cm gas cell. Close C, open B, and slowly open E to fill the line with air. Remove the gas cell and measure its infrared spectrum. If the spectrum is too weak or intense, the concentration of NO$_2$–N$_2$O$_4$ will have to be altered by adding or removing gas. In this case, put the cell back on the line, close E, and slowly open A to evacuate the system. Close A, open C, and measure the pressure. If you think it is too low to give a good spectrum, add more N$_2$O$_4$ by opening D and allowing the N$_2$O$_4$ to warm up as you did earlier. By adjusting the pressure, a good spectrum of the NO$_2$–N$_2$O$_4$ mixture can be obtained. The NO$_2$ spectrum should contain three absorptions ($3N - 6 = 3$), whereas the N$_2$O$_4$ spectrum is somewhat more complex. Using the references at the end of this experiment, assign all of the bands in the spectrum to either NO$_2$ or N$_2$O$_4$. Note that each band is composed of a number of small peaks (fine structure) that represent transitions to different rotational energy levels.

Purification of N$_2$O$_4$ for Use in the Preparation of Cu(NO$_3$)$_2$ in Experiment 11

This distillation should be carried out immediately before the N$_2$O$_4$ is to be used. Fill the drying column with fresh P$_4$O$_{10}$ and assemble the distillation apparatus as shown in Figure 10–3; note that most of the components were used in the apparatus in Figure 10–1. Extend the glass tubing to the bottom of tube F with Tygon tubing. Attach an empty tube at F, immerse the U-tube in a $-78°$C Dewar flask, and flush the system for about 10 minutes with dry N$_2$. Allow the N$_2$O$_4$ in the original trap (Figure 10–1) to melt, and with the N$_2$ still flowing, replace the empty tube at F with the collection tube containing the N$_2$O$_4$. Adjust the N$_2$ flow to approximately 4 bubbles per second through the bubbler to maintain a relatively slow rate of N$_2$O$_4$ evaporation. Cool the collection trap at the extreme right of Figure 10–3 to $-78°$C and collect the solid N$_2$O$_4$. To maintain a reasonable but collectable rate of distillation, adjust the N$_2$ flow rate and warm the N$_2$O$_4$ (in F) with a beaker of warm water. When the distillation is complete, stopper the N$_2$O$_4$ collection trap and store the compound in a freezer or $-78°$C bath. Preferably, it should be used immediately. When the experiment is complete, place the open tube containing excess N$_2$O$_4$ in the hood at room temperature and allow it to evaporate. Again be certain to clean the glassware immediately to avoid frozen ground glass joints.

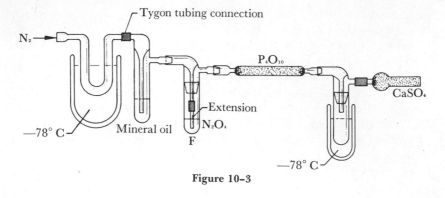

Figure 10–3

REPORT

Include the following:
1. Approximate percentage yield of N_2O_4.
2. Infrared spectrum of gaseous NO_2–N_2O_4, with assignments of the absorptions to NO_2 and N_2O_4.

QUESTIONS

1. What impurities are removed from the N_2O_4 in the final distillation? Are these impurities visible in the infrared spectrum?
2. What is the purpose of the P_4O_{10} drying column in both the preparation and distillation of N_2O_4?
3. If some NO were formed in the preparation, at what stage would it be removed from the N_2O_4?
4. Give Lewis diagrams (dot formulas) for gaseous and solid N_2O_5, NO_2, and N_2O_4.
5. Give the formula of at least one other ionic substance, besides N_2O_5, that contains the nitronium, NO_2^+, ion.
6. What is the structure of P_4O_{10}?
7. Write an equation for the reaction of P_4O_{10} with excess water. A partially hydrated form of P_4O_{10} is given on page 90 as "HPO_3." What structure(s) does HPO_3 have?
8. How does fuming HNO_3 differ from concentrated HNO_3?
9. Write an equation for the reaction of N_2O_4 with water.

INDEPENDENT STUDIES

A. Determine the value of the equilibrium constant for the gas-phase dissociation, $N_2O_4 \rightleftarrows 2NO_2$, using UV-visible spectroscopy. (L. Harris and K. L. Churney, *J. Chem. Phys.*, 47, 1703 (1967).)
B. Determine the value of the equilibrium constant for the dissociation, $N_2O_4 \rightleftarrows 2NO_2$, in cyclohexane or CCl_4 solution, using proton nmr

spectrometry. (T. F. Redmond and B. B. Wayland, *J. Phys. Chem.*, *72*, 1626 (1968).)

C. Record the infrared spectrum of SO_2 and assign absorptions to the correct vibrational modes. (A. G. Briggs, *J. Chem. Educ.*, *47*, 391 (1970).)

D. Prepare nitrosyl chloride (ClNO) and record its infrared spectrum. (C. T. Ratcliffe and J. M. Shreeve, *Inorganic Syntheses, Vol. XI,* McGraw-Hill, New York, 1968, p. 194.)

REFERENCES

N_2O_4

C. F. Bell, *Syntheses and Physical Studies of Inorganic Compounds,* Pergamon Press, New York, 1972, Chap. 4. A summary of physical studies of N_2O_4.

G. Herzberg, *Molecular Spectra and Molecular Structure, II. Infrared and Raman Spectra of Polyatomic Molecules,* D. Van Nostrand Co., Princeton, New Jersey, 1945, pp. 184, 284. Infrared absorptions of N_2O_4 and NO_2.

K. Nakamoto, *Infrared Spectra of Inorganic and Coordination Compounds,* 2nd ed., John Wiley and Sons, New York, 1970. Infrared spectra of NO_2 and N_2O_4.

A. Pedler and F. H. Pollard, *Inorganic Syntheses, Vol. V,* McGraw-Hill, New York, 1957, p. 87. Preparation of N_2O_4.

H. Siebert, *Anwendungen der Schwingungsspektroskopie in der anorganischen Chemie,* Springer-Verlag, Berlin, 1966. Infrared absorptions of NO_2 and N_2O_4.

H. H. Sisler, *J. Chem. Educ., 34,* 555 (1957). Chemistry of N_2O_4.

H. H. Sisler, *Chemistry in Non-aqueous Solvents,* Reinhold Publishing Corp., New York, 1961, Chapter 4. Paperback. Solvent and chemical properties of N_2O_4.

Furnace Techniques

See references at the end of Experiment 3.

Vacuum Line Techniques

See Experiment 19 and Introduction.

Anhydrous Cu(NO₃)₂

Note: The N_2O_4 required in this experiment may be obtained commercially or prepared by the method in Experiment 10.

The nitrate ion, NO_3^-, plays a prominent role in inorganic chemistry. It may be found as the free ion or as a ligand coordinated to a metal ion, as in $Co(NH_3)_5(NO_3)^{2+}$ or $Fe(NO_3)_4^-$. As the ion it has a symmetrical planar structure analogous to that of the isoelectronic CO_3^{2-} ion. In solid $NaNO_3$, it has the following dimensions:

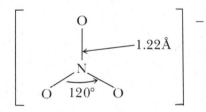

Such ionic nitrates, MNO_3, of the alkali metals (sodium through cesium) may be prepared by simply reacting the metal or metal oxide with HNO_3. Crystallization from water yields the anhydrous metal nitrate. On the other hand, hydrated nitrate salts (e.g., $Ca(NO_3)_2 \cdot 4H_2O$ and $Sr(NO_3)_2 \cdot 2H_2O$) of the +2 alkaline earth metal ions precipitate from aqueous solutions under similar conditions.

In most cases, these complexes may be dehydrated by heating to give, e.g., anhydrous $Ca(NO_3)_2$ and $Sr(NO_3)_2$. The small Be^{2+} and Mg^{2+} ions, however, have a strong affinity for H_2O, and their hydrated nitrate salts, $Be(NO_3)_2 \cdot 4H_2O$ and $Mg(NO_3)_2 \cdot 6H_2O$, cannot be dehydrated by heating. Instead, these compounds decompose with the loss of HNO_3 and the formation of metal oxides or hydroxides. Hence, the synthesis of anhydrous metal nitrates from hydrated metal salts is not possible when the metal ion coordinates strongly with H_2O. This consideration suggests that the strong coordination tendencies of transition metal ions will make dehydration of their hydrated nitrate salts very difficult. This is true, and in general, anhydrous nitrate salts of these metals are most easily prepared by methods involving the rigorous exclusion of water. The problems associated with the preparation of anhydrous metal nitrates are similar in many ways to those encountered in the preparation of anhydrous metal halides described in Experiment 3.

A highly reactive reagent for generating anhydrous metal nitrates is dinitrogen pentoxide, N_2O_5, the anhydride of HNO_3. In the gas phase, it has the molecular structure

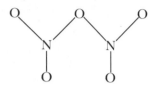

but consists of NO_2^+ and NO_3^- ions in the solid state. The NO_2^+ ion is linear and isoelectronic with CO_2. The nature of the reactions of N_2O_5 in solvents suggests that it is these ions that are involved. For example, the preparation of anhydrous $Sn(NO_3)_4$ occurs by the simple metathesis reaction:

$$SnCl_4 + 4NO_2^+NO_3^- \rightarrow Sn(NO_3)_4 + 4NO_2Cl(g)$$

The product is very volatile and may be purified by sublimation at 40°C under vacuum. Infrared evidence suggests that each NO_3^- group is bidentate

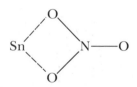

and that the Sn is 8-coordinated. An x-ray structural investigation has demonstrated the presence of this type of coordination in $Ti(NO_3)_4$.

Although N_2O_5 is very effective in preparing anhydrous metal nitrates, it is difficult to prepare and it slowly decomposes to N_2O_4 and O_2 even at 0°. These difficulties have led researchers to shift their attention to more conveniently handled nitrating agents such as N_2O_4. The latter is commercially available and may be handled as a liquid (m.p. −12.3°, and b.p. 21.3°C). It has the usual disadvantages of a volatile toxic liquid, and it is also very hygroscopic and should be protected from atmospheric moisture at all times. It must always be handled in the hood. In the solid and gas phases, N_2O_4 exists as the following planar symmetrical molecule:

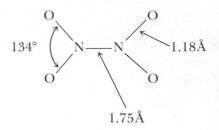

The unusually long N—N bond (as compared with hydrazine, H_2NNH_2, where the N—N distance is 1.47Å) is consistent with its tendency to dissociate into two molecules of NO_2. Thus, the colorless N_2O_4 in the liquid and gaseous phases is in equilibrium with brown paramagnetic NO_2. In liquid N_2O_4 relatively little dissociation to NO_2 occurs; even at the

$$N_2O_4 \rightleftharpoons 2NO_2$$

boiling point it contains only 0.1 per cent NO_2. Dissociation increases in the gas phase such that the mixture contains 90 per cent NO_2 at 100°C.

Another equilibrium of importance in N_2O_4 chemistry is that involving autoionization to nitrosonium, NO^+, and nitrate ions.

$$N_2O_4 \rightleftharpoons NO^+ + NO_3^-$$

Although conductance measurements on liquid N_2O_4 indicate that the extent of this dissociation is very small, the chemical reactivity of N_2O_4 suggests that it frequently reacts as if NO^+ and NO_3^- ions were present. An example is the reaction of N_2O_4 with $ZnCl_2$:

$$2N_2O_4 + ZnCl_2 \rightarrow 2ClNO + Zn(NO_3)_2$$
$$(NO^+NO_3^-)$$

The involvement of NO^+ and NO_3^- ions in this reaction is further supported by the fact that the addition of donor solvents such as diethyl ether, ethyl acetate, or benzyl cyanide greatly accelerate the rate of this and similar reactions. It is known that such donor solvents, D, greatly increase the ionic conductivity of liquid N_2O_4, presumably by facilitating ionization to NO^+ and NO_3^-.

$$nD + N_2O_4 \rightleftharpoons D_nNO^+ + NO_3^-$$

Solid adducts with formulas of $D \cdot N_2O_4$ or $D_2 \cdot N_2O_4$ have been isolated in some cases.

The preparation of anhydrous $Cu(NO_3)_2$ in this experiment takes advantage of the nitrating properties of liquid N_2O_4 that contains the donor solvent ethyl acetate. Since metallic Cu will be used, the reaction is not a metathesis reaction as occurred with $ZnCl_2$, but instead involves the oxidation of Cu and reduction of NO^+:

$$3N_2O_4 + Cu \xrightarrow{\text{ethyl acetate}} 2NO + Cu(NO_3)_2 \cdot N_2O_4$$
$$(NO^+NO_3^-)$$

Nitric oxide bubbles out of solution with simultaneous precipitation of $Cu(NO_3)_2 \cdot N_2O_4$. Infrared spectral and conductivity data suggest that $Cu(NO_3)_2 \cdot N_2O_4$ exists as the ionic complex $NO^+[Cu(NO_3)_3^-]$ in solution. An x-ray structural investigation of the solid shows that it consists of NO^+ cations and a polymeric anion in which NO_3^- groups bridge the copper atoms. Despite its structure, one molecule of N_2O_4 per copper atom can be removed from the solid by heating it to 120°C under vacuum. The resulting blue $Cu(NO_3)_2$ may be sublimed at 200°C under vacuum. Electron diffraction and electron spin resonance studies of $Cu(NO_3)_2$ show that it has the following square planar geometry in the gas phase:

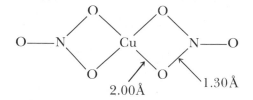

In the solid state, the structure changes. The nitrate groups are no longer bidentate toward one Cu atom, but each NO_3^- group bridges between two Cu atoms to give a polymeric structure. Such changes in

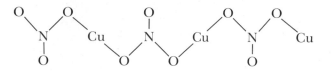

structure from the gaseous to the solid state are not uncommon in many types of inorganic compounds. Anhydrous $Cu(NO_3)_2$ is very hygroscopic and will react immediately with moisture in the air to form the more common hydrated salt, $Cu(NO_3)_2 \cdot 3H_2O$.

EXPERIMENTAL PROCEDURE

NO_2–N_2O_4 is as toxic as carbon monoxide (CO) and hydrogen cyanide (HCN), and *must always be handled in an efficient hood.* The exceedingly hygroscopic nature of anhydrous $Cu(NO_3)_2$ requires that all equipment and chemicals be rigorously dry. The yield of the product will largely depend upon the care exercised in excluding moisture. The reaction will be conducted in the all-glass apparatus shown in Figure 11–1. Rubber stoppers may not be used because they deteriorate in N_2O_4. Fill the drying tube with P_4O_{10}, and keep the apparatus assembled (lubricate joints with silicone grease) as in Figure 11–1 except when reactants are being added.

Introduce a magnetic stirring bar and 2 ml of a good grade of dry ethyl acetate into the tube in the apparatus. (The ethyl acetate should be dried by distilling it from P_4O_{10}. It may be stored over Linde type 4A molecular sieves. Proper drying of the ethyl acetate is important.) Then add an estimated 2 ml of N_2O_4 (b.p., 21.3°). Replace the drying tube and cool the solution in an ice bath. The N_2O_4 may be prepared as in Experiment 10 or obtained commercially in a compressed gas cylinder. A lecture bottle gas cylinder with a needle valve may simply be inverted; the N_2O_4 can be poured into the tube by controlling the flow with the needle valve. Larger cylinders of N_2O_4 are frequently of the type shown in Figure 11–2; these are equipped with a full length eductor tube and valves for discharging either gaseous or liquid N_2O_4.

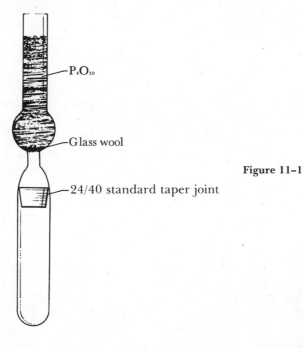

— P_4O_{10}

— Glass wool

Figure 11–1

— 24/40 standard taper joint

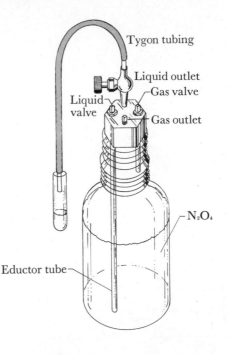

Tygon tubing

Liquid outlet

Gas valve

Liquid valve

Gas outlet

N₂O₄

Eductor tube

Figure 11–2

The liquid valve on the cylinder is opened with a square wrench, and the attached needle valve is used to adjust the flow of the liquid into the flask. Be sure to protect the contents of the flask with the P_4O_{10} drying tube. Remove the drying tube long enough to add 1.0 g of powdered copper metal to the solution. Nitric oxide is evolved, and $Cu(NO_3)_2 \cdot N_2O_4$ begins to precipitate. After stirring magnetically at 0° for about 15 minutes, add 2 additional milliliters of N_2O_4 to the semisolid mixture. Stir for approximately 45 minutes more at 0°C.

Then quickly remove the drying tube and insert the sublimation probe as shown in Figure 11–3. Attach the sublimation apparatus to the utility vacuum line and evaporate the mixture to dryness under a dynamic vacuum. Evaporation proceeds more rapidly if the reaction mixture is warmed with a beaker of warm water. The solid that remains is the adduct, $Cu(NO_3)_2 \cdot N_2O_4$, and unreacted copper. Remove the N_2O_4 from $Cu(NO_3)_2 \cdot N_2O_4$ by slowly heating the solid in a silicone fluid bath to 85°, still under a dynamic vacuum. Hold the temperature near 85° until all of the N_2O_4 has been discharged (~2 hours). This can be checked by momentarily closing the stopcock at the sublimer; if N_2O_4 is still being evolved, the tube will fill with brown NO_2 gas. When NO_2 is no longer being evolved, raise the temperature of the silicone bath to 120° to drive off any remaining N_2O_4. Determine when this is complete by noting the brown NO_2 gas as before. (The removal of N_2O_4 from $Cu(NO_3)_2 \cdot N_2O_4$ must be done slowly, as described, in order to avoid a

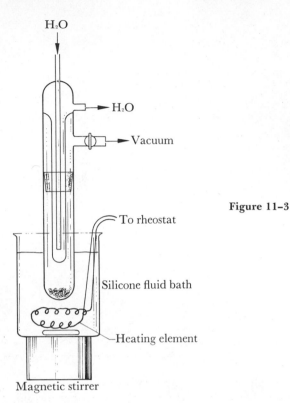

H_2O

H_2O

Vacuum

Figure 11–3

To rheostat

Silicone fluid bath

Heating element

Magnetic stirrer

sudden surge in pressure, which may separate the sublimation tube from the sublimation unit. *This procedure, as well as the following sublimation, should be done behind a safety shield.*

With the sample still under a dynamic vacuum, turn on the water flow through the cold finger probe, and raise the temperature of the bath to 210°C. When the bath has reached 210°, close the stopcock at the sublimer. From time to time, this stopcock should be opened briefly to ensure a good vacuum in the sublimation tube.

Anhydrous $Cu(NO_3)_2$ will sublime onto the probe and perhaps onto the cooler areas of the sublimer tube. (On humid days, moisture will condense on the outside of the sublimer and drop into the hot silicone bath. Since water is denser than the oil, drops of water go to the bottom of the bath where they vaporize explosively, spewing hot fluid out of the bath. Wrap an absorbent cloth around the sublimer to prevent water from dropping into the bath. *This is important.*)

To weigh and determine the melting point of the blue $Cu(NO_3)_2$, it will have to be handled in a dry bag. First carefully empty the water out of the cold finger probe of the sublimer and rinse it with acetone. After drying it in a stream of air, put the sublimation apparatus, together with a preweighed bottle and cover, several capillary tubes sealed at

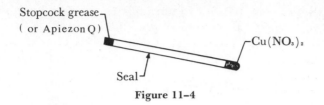

Figure 11–4

one end, a spatula, a tube of stopcock grease (or Apiezon Q), and a small dry mortar and pestle into the dry bag (see Experiment 9, p. 85). With the entrance port open, flush the bag with a brisk stream of nitrogen gas for about 5 minutes. Then roll up the entrance port and clamp it closed. Allow the nitrogen to fill the bag to a convenient and workable pressure; then turn off the nitrogen flow. Powder your hands with talcum powder and insert them into the gloves in the bag.

Open the stopcock on the sublimer and remove the tube containing the unsublimed residue. Scrape the $Cu(NO_3)_2$ off the cold finger into the mortar. Transfer most of the product to the weighing bottle and cap it tightly. Powder the remaining $Cu(NO_3)_2$ in the mortar and pestle and, by dipping the open end of a capillary tube into the powder, fill the tube to a height of about 0.5 cm with the $Cu(NO_3)_2$. After tapping the sample to the bottom, close the open end of the tube to the air by plugging it with as little stopcock grease (or Apiezon Q) as is necessary. Prepare one or two other capillary tubes similarly. Now open the entrance port of the dry bag; seal the capillary tubes with a small flame roughly in the position shown in Figure 11–4. Determine the melting point of the anhydrous $Cu(NO_3)_2$ and also of $Cu(NO_3)_2 \cdot 3H_2O$. Weigh the weighing bottle and calculate the percentage yield. Remove a small amount of anhydrous $Cu(NO_3)_2$ and observe its reaction with air.

REPORT

Include the following:
1. Percentage yield of $Cu(NO_3)_2$.
2. Melting or decomposition temperatures of $Cu(NO_3)_2$ and $Cu(NO_3)_2 \cdot 3H_2O$.
3. Observations and explanation of $Cu(NO_3)_2$ reaction with air.

QUESTIONS

1. Draw a probable structure for $Cu(NO_3)_2 \cdot 3H_2O$.
2. Draw a probable structure for the ether adduct, $N_2O_4 \cdot 2(C_2H_5)_2O$.
3. Rationalize the extent of hydration of the following series of alkaline earth nitrates: $Mg(NO_3)_2 \cdot 6H_2O$, $Ca(NO_3)_2 \cdot 4H_2O$, $Sr(NO_3)_2 \cdot 2H_2O$, and $Ba(NO_3)_2$.

4. What types of compounds can normally be purified by sublimation? Would you expect $Cu(NO_3)_2 \cdot 3H_2O$ to sublime?

5. A P_4O_{10} drying tube is used to protect the $Cu-N_2O_4$ reaction mixture from atmospheric moisture. Why cannot the reaction tube simply be stoppered instead?

6. (a) How would you determine experimentally whether or not $Cu(NO_3)_2$ dissolved in nitromethane was ionic? Would you expect it to be ionic? (b) Would you expect $Cu(NO_3)_2$ to be ionic in H_2O solvent? Explain.

7. Draw structural formulas for the following:
   (a) NO_3^- and CO_3^{2-}
   (b) NO_2Cl and $ClNO$
   (c) solid N_2O_5

8. Draw a Lewis dot formula for $Cu(NO_3)_2$.

9. Complete the following equations:
   (a) $Mg(NO_3)_2 \cdot 6H_2O \xrightarrow{\Delta}$
   (b) $Na + \text{excess } N_2O_4 \rightarrow$

INDEPENDENT STUDIES

A. Measure the mass spectrum of $Cu(NO_3)_2$ and assign m/e values to the ion fragments. See Experiment 14. (R. F. Porter, R. C. Schoonmaker and C. C. Addison, *Proc. Chem. Soc.*, 11 (1959).)

B. Prepare and characterize $Cu(NO_3)_2 \cdot N_2O_4$. (C. C. Addison and B. J. Hathaway, *J. Chem. Soc.*, 1468 (1960).)

C. Prepare and characterize $Co(NO_3)_2 \cdot 2N_2O_4$, $Co(NO_3)_2 \cdot N_2O_4$, and/or $Co(NO_3)_2$. (C. C. Addison and D. Sutton, *J. Chem. Soc.*, 5553 (1964).)

D. Prepare and characterize $Zn(NO_3)_2$. (C. C. Addison, J. Lewis, and R. Thompson, *J. Chem. Soc.*, 2829 (1951); B. O. Field and C. J. Hardy, *J. Chem. Soc.*, 4428 (1964).)

E. Prepare $Cu(N\equiv C-CH_3)_2(NO_3)_2$ from $Cu(NO_3)_2$ and CH_3CN. (B. Duffin, *Acta Cryst.*, 24B, 396 (1968).)

REFERENCES

$Cu(NO_3)_2$

C. C. Addison and B. J. Hathaway, *J. Chem. Soc.*, 3099 (1958). Preparation and characterization of $Cu(NO_3)_2$.

Anhydrous Metal Nitrates

C. C. Addison and N. Logan in *Preparative Inorganic Reactions, Vol. 1,* W. L. Jolly, ed., Interscience Publishers, New York, 1964, p. 141. Methods of preparing anhydrous metal nitrates.

C. C. Addison and N. Logan in *Advances in Inorganic Chemistry and Radiochemistry, Vol. 6,* H. J. Emeleus and A. G. Sharpe, eds., Academic Press, New York, 1964, p. 72. Survey of anhydrous metal nitrates with preparations and discussions of structures.

C. C. Addison and D. Sutton, in *Progress in Inorganic Chemistry, Vol. 8,* F. A. Cotton, ed., Interscience Publishers, New York, 1967, p. 195. An extensive survey of complexes containing the nitrate ion.

C. C. Addison, N. Logan, C. D. Garner, and S. C. Wallwork, *Quart. Rev., 25,* 289 (1971). Structural aspects of coordinated nitrate groups.

L. J. Blackwell, T. J. King, and A. Morris, *J. C. S. Chem. Comm.,* 644 (1973). Solid state structure of $Cu(NO_3)_2 \cdot N_2O_4$.

Complex Ion Composition by Job's Method

Frequently it is possible to detect the interaction of two molecules in solution without being able to isolate a stable compound. For example, benzene and iodine, I_2, interact in carbon tetrachloride

$$C_6H_6 + I_2 \rightleftharpoons C_6H_6 \cdot I_2$$

to form a highly colored 1:1 adduct that is too unstable for isolation. The presence of an adduct is demonstrated by the intense color of these solutions, yet its composition is uncertain until it has been established that only one molecule of C_6H_6 combines with one molecule of I_2.

Coordination chemistry has depended heavily on studies of complexes that may be identified in solution without actually being isolated. The interaction, for example, of Ni^{2+} with NH_3 in water produces complexes of the compositions $Ni(NH_3)(OH_2)_5^{2+}$, $Ni(NH_3)_2(OH_2)_4^{2+}$, $Ni(NH_3)_3(OH_2)_3^{2+}$, $Ni(NH_3)_4(OH_2)_2^{2+}$, $Ni(NH_3)_5(OH_2)^{2+}$, and $Ni(NH_3)_6^{2+}$. Of this series, only $Ni(NH_3)_6^{2+}$ has actually been isolated; yet spectrophotometric and potentiometric investigations leave little doubt concerning the existence of the others in solution. The fact that the complexes have not been isolated does not always imply that the interactions are weak. Indeed, bond formation between transition metal ions and ligands is highly exothermic. For other reasons, however, it is frequently not possible to crystallize from solution all of the species that may be present in solution. Their composition must then be established by other techniques.

The procedure to be used in determining the solution composition of Ni^{2+}–ethylenediamine complexes in this experiment is known as the method of continuous variations or Job's method. In the general case, it is concerned with evaluating n for the equilibrium,

$$Z + nL \rightleftharpoons ZL_n \tag{1}$$

In the present experiment, Z is Ni^{2+} and L is the ligand ethylenediamine (en). Both Ni^{2+} and the product $Ni(en)_n^{2+}$ complexes have absorptions in the visible region of the light spectrum, but their spectra are different.

Experimentally, the intensity of absorption at a given wavelength of a series of solutions containing varying amounts of Ni^{2+} and en is measured. This absorbance is related to the concentration of $Ni(en)_n^{2+}$ in solution. These solutions are prepared with the restriction that the sum of the concentrations of Ni^{2+} and en be the same in all solutions. In the case where the equilibrium constant for reaction (1) is very large (i.e., the equilibrium lies far toward the right), it is clear that the intensity of the $Ni(en)_n^{2+}$ absorption will be greatest when the en concentration in solution is exactly n times greater than that of Ni^{2+}. As will be shown later, this is also true when the equilibrium does not lie far to the right. Sufficient concentrations of ZL_n must, however, be produced so that accurate absorbance measurements may be obtained on the solutions. It is therefore possible to determine n and the composition of $Ni(en)_n^{2+}$ by knowing the ratio of en to Ni^{2+} in the solution that contains a maximum absorbance for $Ni(en)_n^{2+}$.

For the Ni^{2+}–en system, a series of complexes, $Ni(en)^{2+}$, $Ni(en)_2^{2+}$, $Ni(en)_3^{2+}$, $Ni(en)_4^{2+}$, and so forth, are possible. The purpose of this experiment is to determine which of these species are actually present in solution. The possible equilibria involved follow:

$$Ni^{2+} + en \overset{K_1}{\rightleftharpoons} Ni(en)^{2+}$$

$$Ni(en)^{2+} + en \overset{K_2}{\rightleftharpoons} Ni(en)_2^{2+}$$

$$Ni(en)_2^{2+} + en \overset{K_3}{\rightleftharpoons} Ni(en)_3^{2+}$$

$$Ni(en)_3^{2+} + en \overset{K_4}{\rightleftharpoons} Ni(en)_4^{2+}, \text{ etc.}$$

The values of the equilibrium constants will determine which species will predominate in solution. Thus, if K_2 is so much larger than K_1 that virtually no $Ni(en)^{2+}$ is present in solution, the Job procedure will identify $Ni(en)_2^{2+}$ without giving any evidence of $Ni(en)^{2+}$. Likewise, the relative size of K_3 with respect to K_1 and K_2 will determine what species are characterized. One of the limitations of the method of continuous variations is the requirement that only one equilibrium of the type in equation (1) be present in a solution of Z and L. That is, it will give non-integral values of n if in addition to Z, L, and ZL_n another complex, ZL_{n+1}, is also present. For the Ni^{2+}–en reactions, this means that only two complexes [Ni^{2+} and $Ni(en)^{2+}$, or $Ni(en)^{2+}$ and $Ni(en)_2^{2+}$, and so forth] can be present in any given solution. This will be true if K_1, K_2, and K_3 are greatly different. From potentiometric studies of the interactions of Ni^{2+} and en, the values of K_1, K_2, and K_3 have been evaluated (see Experiment 13), and they in fact are separated by large factors. Although Job's method allows the determination of complex compositions in this system, misleading or erroneous results might be obtained in a system where the K values for successive equilibria are unknown. For this reason, the method of continuous variations is limited to relatively simple systems.

Theory

The purpose of this section is to prove that the value of n in ZL_n of equation (1) may be determined from spectrophotometric absorbance measurements on a series of solutions containing varying amounts of Z and L yet having the same total concentration of Z plus L. If the absorbance at a given wavelength of each solution is plotted *versus* the mole fraction, X, of L in the solution, the maximum absorbance will occur at a mole fraction which corresponds to the composition of ZL_n. Hence, n is determined.

Assume that substances Z and L react according to equation (1). Equimolar solutions of Z and L, each of M moles per liter concentration, are mixed in varying amounts so that the total concentration (Z and L) is M. A series of these solutions may be prepared by the addition of X liters of L to $(1-X)$ liters of Z (where $X<1$). The concentrations of Z, L, and ZL_n at equilibrium in these solutions are designated C_1, C_2, and C_3, respectively. Thus, for any solution the concentrations are expressed as follows:

$$C_1 = M(1-X) - C_3 \qquad (2)$$
$$C_2 = MX - nC_3 \qquad (3)$$
$$C_3 = KC_1C_2{}^n \qquad (4)$$

where K is the equilibrium constant for reaction (1). The condition for a maximum in the curve of C_3 plotted versus X is that

$$\frac{dC_3}{dX} = 0 \qquad (5)$$

Differentiation of equations (4), (3), and (2) with respect to X and combination of the three resulting differential equations with equations (2), (3), and (5) gives

$$n = \frac{X}{1-X} \qquad (6)$$

Therefore, from the value of X for which C_3 is a maximum, n may be calculated from equation (6).

Now it remains to be shown that a maximum in the absorbance at a given wavelength of light when X is varied coincides with the maximum of C_3. From the Beer-Lambert Law,

$$A = \epsilon Cl \tag{7}$$

where A = absorbance, ϵ = molar extinction coefficient, C = molar concentration, and l = length of the light path or thickness of the cell.

The extinction coefficients of Z, L, and ZL_n at a given wavelength are designated ϵ_1, ϵ_2, and ϵ_3, respectively. Since the absorbance of a solution is the sum of the absorbances at that wavelength of the contained species, the measured absorbance, A_{meas}, is

$$A_{meas} = (\epsilon_1 C_1 + \epsilon_2 C_2 + \epsilon_3 C_3)l \tag{8}$$

If there is no interaction between Z and L, i.e., $C_3 = 0$, the absorbance, A_{Z+L}, would be

$$A_{Z+L} = \left[\epsilon_1 M(1-X) + \epsilon_2 MX \right] l \tag{9}$$

where M is still the molar concentration of the Z and L solutions.

The difference between A_{meas} and A_{Z+L} is designated Y.

$$Y = \left[\epsilon_1 C_1 + \epsilon_2 C_2 + \epsilon_3 C_3 - \epsilon_1 M(1-X) - \epsilon_2 MX \right] l \tag{10}$$

By differentiation of equation (10) with respect to X it can be shown that Y is a maximum when C_3 is a maximum if $\epsilon_3 > \epsilon_1$, or a minimum when C_3 is a maximum if $\epsilon_3 < \epsilon_1$.

In this experiment Ni^{2+} corresponds to Z, and ethylenediamine (en) to L. Since en has no absorptions in the region of the spectrum under study, ϵ_2 will always be zero. The cell to be used in the measurements has a thickness of 1 cm. Under these conditions equation (10) reduces to

$$Y = A_{meas} - (1-X)A_Z \tag{11}$$

where A_Z is the absorbance of the pure M molar Ni^{2+} solution. To evaluate n in $Ni(en)_n^{2+}$, a plot of Y (ordinate) versus X (abscissa) at a given wavelength is made. A maximum in this plot occurs at a certain X mole fraction. From this value of X, n may be calculated from equation (6). Because more than one complex composition will be examined in this experiment, plots will be constructed from data taken at several wavelengths corresponding to different $N(en)_n^{2+}$ complexes.

EXPERIMENTAL PROCEDURE

Prepare 100 ml of the following aqueous solutions:
0.4 M nickel sulfate, $NiSO_4 \cdot 6H_2O$
0.4 M ethylenediamine

By mixing these solutions, prepare mixtures having a total volume of 10 ml in which the mole fraction of ethylenediamine, X, is 0.3, 0.4, 0.5, 0.6, 0.7, 0.8, and 0.9. Determine the absorbances of each of these solutions and of the pure 0.4 M nickel sulfate solution at the following wavelengths: 530, 545, 578, 622, and 640 nm. If a scanning visible spectrophotometer is available, the most convenient method of determining the absorbances at the various wavelengths is to record the spectrum of each solution from approximately 500 to 650 nm. The spectra of all solutions can be run on the same piece of chart paper. For a fixed wavelength spectrophotometer, measure the absorbances by manually setting the instrument at each of the five wavelengths. (For comments on ultraviolet visible spectrophotometers, see Experiment 2, p. 29.)

Calculate Y values (from equation 11) at each wavelength for the entire series of solutions, and make plots of Y *versus* X for each of the five wavelengths. From the value of X at each of the five maxima, the values of $\hat{n}$ for the $Ni(en)_n^{2+}$ complexes may be calculated from equation (6).

REPORT

Include the following:
1. Spectra or absorbance measurements on the solutions.
2. Plots of Y *versus* X for each of the five wavelengths.
3. Calculated values of n and formulas of Ni^{2+} complexes present in solution.

QUESTIONS

1. Draw structures for the $Ni(en)_n^{2+}$ complexes in solution.
2. From the observed changes in the spectra and a knowledge of the complexes present in the solutions, what can be said about the relative ligand field strengths of en and H_2O?
3. Derive equation (6).
4. By differentiation of equation (10), show that Y is a maximum when C_3 is a maximum.
5. Give a specific example of a reaction in which Job's method would be useful.
6. Explain why "it is clear that the intensity of the $Ni(en)_n^{2+}$ absorption will be greatest when the en concentration in solution is exactly n times greater than that of Ni^{2+}." (The statement is taken from page 109, paragraph 1, line 7.)

7. A reaction occurs between $Co(ClO_4)_2$ and $LiNO_3$ in t-butanol to form a complex of the type $Co(NO_3)_n^{(2-n)+}$. The following absorbance data at a specific wavelength were collected on solutions made up in the given concentrations of $Co(ClO_4)_2$ and $LiNO_3$:

$Co(ClO_4)_2$, M	0.100	0.080	0.060	0.050	0.040	0.030	0.020	0.000
$LiNO_3$, M	0.000	0.020	0.040	0.050	0.060	0.070	0.080	0.100
Absorbance	0.250	0.302	0.354	0.385	0.433	0.408	0.329	0.000

   The $LiNO_3$ does not absorb at this wavelength.
   (a) What is the composition of the complex?
   (b) Draw a probable structure for the complex.
8. Consider the general reaction $Z + L \rightleftarrows ZL$. Draw a general plot of Y *versus* X for the case in which the equilibrium constant is very large and favors the formation of ZL. On the same graph, plot Y *versus* X for the case in which the equilibrium constant is not large, i.e., appreciable concentrations of Z, L, and ZL are present in all solution mixtures. Assume that the extinction coefficient of ZL is greater than that of Z, and that L does not absorb. Explain why the two plots differ.

INDEPENDENT STUDIES

A. Using Job's method, determine the compositions of complexes present in solutions of Ni^{2+} and 1,3-diaminopropane.
B. Using Job's method, determine the compositions of complexes present in solutions of Cu^{2+} and iminodiacetic acid, $HN(CH_2CO_2H)_2$. (C. N. Reilley and D. T. Sawyer, *Experiments for Instrumental Analysis,* McGraw-Hill, New York, 1961, p. 178.)
C. Using Job's method, determine the compositions of complexes present in solutions of Ni^{2+} and diethylenetriamine, $NH_2CH_2CH_2$-$NHCH_2CH_2NH_2$.
D. Prepare and isolate a salt of $Ni(en)_3^{2+}$. Compare its visible spectrum with spectra you obtained in this experiment. Possibly further characterize the compound by its magnetic susceptibility, conductivity, infrared spectrum, etc. (K. H. Pearson, W. R. Howell, Jr., P. E. Reinbold, and S. Kirschner, *Inorganic Syntheses, Vol. 14,* McGraw-Hill, New York, 1973, p. 61.)

REFERENCES

$Ni(en)_n^{2+}$

W. C. Vosburgh and G. R. Cooper, *J. Am. Chem. Soc., 63,* 437 (1941).

Methods to Establish Complex Compositions

M. T. Beck, *Chemistry of Complex Equilibria,* Van Nostrand Reinhold, New York, 1970, p. 86.

M. M. Jones, *Elementary Coordination Chemistry,* Prentice-Hall, Englewood Cliffs, New Jersey, 1964, Chapter 8. A brief survey of methods available for determining complex composition and their formation constants.

L. I. Katzin and E. L. Gebert, *J. Am. Chem. Soc., 72,* 5455 (1950). Applications and limitations of Job's method.

W. Likussar and D. F. Boltz, *Anal. Chem., 43,* 1265 (1971). Use of method of continuous variations for determining stability constants.

F. J. C. Rossotti and H. Rossotti, *The Determination of Stability Constants,* McGraw-Hill, New York, 1961, Chapter 3. A detailed treatment of methods of evaluating metal complex formation constants.

K. O. Watkins and M. M. Jones, *J. Inorg. Nucl. Chem., 24,* 1235 (1962).

Stability Constants of Ni(glycinate)$_n^{(2-n)+}$

Although coordination compounds have a variety of geometries, in general the first-row +2 transition metal ions react with water to form octahedral aquo-complexes of the formula $M(OH_2)_6^{2+}$. The addition of another ligand, L^-, to such solutions results in the competition of L^- for coordination sites occupied by H_2O. If M^{2+} is a labile metal ion, such as Zn^{2+}, Cu^{2+}, Ni^{2+}, and Co^{2+}, the equilibrium

$$M(OH_2)_6^{2+} + L^- \rightleftharpoons M(OH_2)_5(L)^+ + H_2O$$

is established very rapidly. The concentrations of each of the species in solution depend upon the ligand properties of L^- and H_2O as well as on their concentrations. Since the concentration of H_2O in dilute solutions is virtually the same regardless of the position of the equilibrium, $[H_2O]$ in the equilibrium expression

$$K = \frac{[M(OH_2)_5(L)^+][OH_2]}{[M(OH_2)_6^{2+}][L^-]}$$

is effectively a constant and is included in the equilibrium constant. It is frequently written as

$$K_1 = \frac{[M(OH_2)_5(L)^+]}{[M(OH_2)_6^{2+}][L^-]}$$

Realizing that H_2O always occupies coordination sites in the complex not filled by L^-, the H_2O ligands are usually omitted from the complex formulas in both the equilibrium reaction and the associated equilibrium constant expression. Thus,

$$M^{2+} + L^- \rightleftharpoons ML^+ \tag{1}$$

$$K_1 = \frac{[ML^+]}{[M^{2+}][L^-]} \tag{2}$$

115

The constant, K_1, is called a stability constant or sometimes a formation constant.

Although true equilibrium constants are functions of the activities, a, of the reactants and products,

$$K_\gamma = \frac{a_{ML^+}}{a_{M^{2+}}a_{L^-}} = \frac{[ML^+]}{[M^{2+}][L^-]} \cdot \frac{\gamma_{ML^+}}{\gamma_{M^{2+}} \cdot \gamma_{L^-}} \tag{3}$$

the evaluation of the activity coefficients, γ, is usually difficult and seldom done. Activity coefficients usually depend upon the ionic strength of the solution, but at infinite dilution $\gamma = 1$, and concentrations and activities become equal. While it would be desirable to study equilibria in very dilute solutions where the γ's are known to be 1, in practice it is not possible. Hence the γ's are usually not known, but since equilibrium constants frequently depend on the ionic strength of the solution, it is at least desirable to maintain a known and constant ionic strength in the solutions under study. Particularly in reactions involving anionic L^- ligands, the value of the equilibrium constant will change with a variation in the ionic strength of the solution.

In most equilibrium studies of metal complex formation, the equilibrium constant will be evaluated from measurements on solutions containing various concentrations of M^{2+} and L^-. Such concentration changes will produce changes in the ionic strength of the solutions, in the γ's, and therefore in K_γ. To keep the ionic strength constant, a large excess of an unreactive ionic salt is added to the solutions. Then any changes in ionic strength owing to changes in the position of equilibrium (1) will be negligible compared with the high concentration of the added salt. The salt is added only to maintain the ionic strength of the medium and should not interact directly with M^{2+} or L^-. Salts such as KNO_3 and $NaClO_4$ have been used extensively because of the low affinity of the NO_3^- and ClO_4^- ions for M^{2+}. In this experiment KNO_3 will be used.

While a large excess of KNO_3 ensures that the ionic strength will be the same for all measurements, the γ's for M^{2+}, L^-, and ML^+ will still be unknown. Since they will not be evaluated, the ratio $\gamma_{ML^+}/\gamma_{M^{2+}}\gamma_{L^-}$ in equation (3) will be an unknown constant. The value of K will therefore be calculated using only the concentrations of ML^+, M^{2+}, and L^-, as in equation (2). Such a K is called a concentration constant and is valid only at the ionic strength used in the determination. The vast majority of stability constants that are reported in the chemical literature are actually concentration constants.

The existence of six coordination sites in many metal complexes means that L^- and H_2O compete for all six sites, and the following equilibria are possible:

$$M^{2+} + L^- \rightleftarrows ML^+, \quad K_1 = \frac{[ML^+]}{[M^{2+}][L^-]}$$

$$ML^+ + L^- \rightleftarrows ML_2, \quad K_2 = \frac{[ML_2]}{[ML^+][L^-]}$$

$$ML_2 + L^- \rightleftarrows ML_3^-, \quad K_3 = \frac{[ML_3^-]}{[ML_2][L^-]}$$

$$ML_3^- + L^- \rightleftarrows ML_4^{2-}, \quad K_4 = \frac{[ML_4^{2-}]}{[ML_3^-][L^-]}$$

$$ML_4^{2-} + L^- \rightleftarrows ML_5^{3-}, \quad K_5 = \frac{[ML_5^{3-}]}{[ML_4^{2-}][L^-]}$$

$$ML_5^{3-} + L^- \rightleftarrows ML_6^{4-}, \quad K_6 = \frac{[ML_6^{4-}]}{[ML_5^{3-}][L^-]}$$

Complexes such as ML_2 and ML_3^-, which represent $ML_2(OH_2)_4$ and $ML_3(OH_2)_3^-$, may exist in *cis* and/or *trans* forms. These equilibrium expressions do not distinguish these isomeric forms. Thus, $[ML_2]$ refers to the total concentration of both isomers of ML_2, and so forth. Usually, for anionic L^- groups the introduction of six L^- ligands into the complex is prevented by the build-up of negative charge on the complex. This is presumably the reason for the rare occurrence of complexes such as MCl_6^{4-}. Neutral L groups such as NH_3 are commonly found to form all the complexes from $M(OH_2)_6^{2+}$ to $M(NH_3)_6^{2+}$. The equilibrium constants, $K_1, K_2 \cdots K_6$, in the series of equilibria are called stepwise stability constants. The equilibrium constants may, for certain purposes (see next section), also be expressed as overall stability constants, β, which are simply products of the stepwise stability constants.

$$\beta_1 = K_1$$

$$\beta_2 = K_1 K_2$$

$$\beta_3 = K_1 K_2 K_3, \text{ etc.}$$

For reactions that involve six separate equilibria as for $Ni^{2+} + NH_3$, the experimental determination of all six K values is a very difficult, but feasible, task. For this and other reasons, the majority of stability constant investigations have been made on reactions of metal ions with chelating ligands. The chelating ligand to be used in this experiment is the naturally occurring amino acid, glycine, $\overset{+}{N}H_3CH_2CO_2^-$. It exists in the zwitterion form, but as the anion it coordinates to metal ions through both the N and an O.

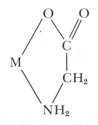

The coordination number of 6 limits the coordination of the metal ion to a maximum of three glycinate ligands. This experiment is concerned with determining the magnitude of the interaction between $NH_2CH_2CO_2^-$ and Ni^{2+}. The equilibrium constants to be determined are shown as follows (the glycinate anion, $NH_2CH_2CO_2^-$, is designated A^-):

$$Ni^{2+} + A^- \rightleftarrows NiA^+ \quad K_1 = \frac{[NiA^+]}{[Ni^{2+}][A^-]} \tag{4}$$

$$NiA^+ + A^- \rightleftarrows NiA_2 \quad K_2 = \frac{[NiA_2]}{[NiA^+][A^-]} \tag{5}$$

$$NiA_2 + A^- \rightleftarrows NiA_3^- \quad K_3 = \frac{[NiA_3^-]}{[NiA_2][A^-]} \tag{6}$$

On the basis of electrostatics, one would expect the affinity of A^- for the complex to decrease as the charge on the complex becomes less positive. It is therefore to be expected that the K's will decrease in the order: $K_1 > K_2 > K_3$.

The experimental determination of the K values for the above equilibria may be carried out by several techniques, but one of the most common is that of measuring with a pH meter the H^+ concentration in a solution containing varying amounts of Ni^{2+} and HA. The H^+ concentration is produced by the ionization of glycine:

$$\overset{+}{N}H_3CH_2CO_2^- \overset{K_a}{\rightleftarrows} H^+ + NH_2CH_2CO_2^-$$
$$HA \qquad\qquad\qquad A^-$$

This equilibrium normally lies far to the left, but addition of Ni^{2+} to the solution results in the release of H^+ depending upon the affinity of Ni^{2+} for the chelating $NH_2CH_2CO_2^-$:

$$Ni^{2+} + HA \overset{K_e}{\rightleftarrows} H^+ + NiA^+$$

A knowledge of the concentrations of Ni^{2+} and HA initially added as well as the H^+ concentration (from pH measurement) at equilibrium allows the calculation of the equilibrium constant, K_e, for this reaction. Since $K_e/K_a = K_1$, the value of K_1 may be evaluated. While this is the general principle for the determination, the formation of three Ni-glycine complexes in the present experiment requires that K_1, K_2, and K_3 all be considered simultaneously; their evaluation is significantly more complicated than just mentioned. This evaluation will be discussed in the next section.

Stability constants have been determined for reactions of a large variety of M^{2+} and L^- groups. Numerous correlations of K values with properties of M^{2+} and L^- have been made. References at the end of the experiment should be consulted for discussions of this area of research. While stability constant studies have been considered to be the domain of the inorganic chemist, recognition of the ligand properties of biologically important molecules has led to extensive investigations of stability constants of metal ions interacting with amino acids, peptides, and nucleic acids. In this and in many other areas, chemists have combined their efforts to solve problems that require a knowledge of both inorganic and biochemistry.

Stepwise Stability Constants of Ni(glycine)$_n$ $^{(2-n)+}$

As noted above, the acidic character of glycine, $\overset{+}{N}H_3CH_2CO_2^-$ (HA), allows stability constants for Ni^{2+} complexation of the glycinate ion, $NH_2CH_2CO_2^-$ (A$^-$), to be evaluated. In order to take advantage of this property, the acid dissociation constant of $\overset{+}{N}H_3CH_2CO_2^-$ must first be determined.

$$HA \rightleftarrows H^+ + A^-$$

$$K_a = \frac{[H^+][A^-]}{[HA]} \tag{7}$$

This is done by measuring $[H^+]$ in a solution of known HA concentration. Before evaluating K_a, let us discuss the measurement of $[H^+]$. A pH meter with a standard glass electrode will be used. Since the pH meter determines the emf of the glass electrode relative to the standard calomel electrode, a pH measurement determines the H^+ activity, a_{H^+}, and not its concentration, $[H^+]$, in solution. Because we want to calculate concentration stability constants, it will be necessary to convert a_{H^+} to $[H^+]$ by using the known activity coefficient of H^+ in a solution having a given ionic strength, μ. The mean activity coefficient, $\gamma_\pm$, of the ions in, e.g., HNO_3, may be calculated from the Davies equation,* an empirical extension of the Debye-Hückel limiting activity equation:

$$-\log \gamma_\pm = \frac{0.50 Z_1 Z_2 \mu^{1/2}}{1 + \mu^{1/2}} - 0.10\mu \tag{8}$$

*C. W. Davies, *J. Chem. Soc.*, 2093 (1938).

where Z_1 and Z_2 are the charges $+1$ and -1 on H^+ and NO_3^-, respectively. The ionic strength of the solution, μ, is given by the usual definition,

$$\mu = \tfrac{1}{2} \sum_i M_i Z_i^2$$

where M_i is the molar concentration of ion i and Z_i is its charge. The ionic strength of the solutions used in this experiment will be determined virtually entirely by the KNO_3 concentration.

By definition,

$$a_{H^+} = \gamma_\pm [H^+]$$

Rearranging and substituting $pH = -\log a_{H^+}$ gives

$$\log [H^+] = -pH - \log \gamma_\pm \qquad (9)$$

Hydroxide ion concentrations may also be calculated from pH measurements by making use of the autoionization constant, $K_w = [H^+][OH^-]$, of water ($K_w = 1.615 \times 10^{-14}$ at 25.0°C and 0.10 M ionic strength):

$$\log [OH^-] = pH - pK_w + \log \gamma_\pm \qquad (10)$$

Hence it is possible to evaluate $[H^+]$ and $[OH^-]$ for use in calculating concentration equilibrium constants from pH measurements.

Returning to the problem of evaluating K_a for glycine, a solution of glycine in 0.10 M KNO_3 is titrated with NaOH. After each addition of NaOH, the pH of the solution is measured. In such solutions, several conditions must hold.

1. The concentration of positive charge must equal that of negative charge, i.e.,

$$[H^+] + [Na^+] = [OH^-] + [A^-]$$

2. The total glycine concentration, A_{tot}, of the prepared solution must be

$$A_{tot} = [HA] + [A^-]$$

3. The expression for the acid dissociation constant, equation (7), pertains.

Combination of these three expressions yields the equation for the pK_a, $-\log K_a$, for glycine:

$$pK_a = -\log [H^+] + \log \left\{ \frac{A_{tot} - ([Na^+] + [H^+] - [OH^-])}{[Na^+] + [H^+] - [OH^-]} \right\} \qquad (11)$$

Since all quantities on the right-hand side of this expression either are known or may be calculated from equations (9) and (10), the pK_a of glycine may be evaluated.

Now the method for determining the stepwise stability constants, K_1, K_2, and K_3, for the reaction of Ni^{2+} with A^- may be introduced. Experimentally a solution containing 1 millimole (mmole) of Ni^{2+} (as $NiCl_2 \cdot 6H_2O$) and 1 mmole of H^+ (as HNO_3) will be titrated with a solution of sodium glycinate, $NH_2CH_2CO_2^-Na^+$, prepared by neutralizing glycine with NaOH. This will give a solution containing an equilibrium mixture of H^+, OH^-, Na^+, HA, A^-, Ni^{2+}, NiA^+, NiA_2, and NiA_3^-. From pH measurements and a knowledge of the quantities of Ni^{2+}, H^+, HA, and NaOH originally added, it is possible to calculate the stability constants.

The method to be used was largely developed by J. Bjerrum.* To facilitate the determination of the K's, a function, $\bar{n}$, is defined as the average number of ligand molecules bound per metal ion. For the present system,

$$\bar{n} = \frac{\text{moles of bound } A^-}{\text{total moles of } Ni^{2+}} = \frac{[NiA^+] + 2[NiA_2] + 3[NiA_3^-]}{[Ni^{2+}] + [NiA^+] + [NiA_2] + [NiA_3^-]}$$

Substituting from equations (4), (5), and (6),

$$\bar{n} = \frac{K_1[A^-] + 2K_1K_2[A^-]^2 + 3K_1K_2K_3[A^-]^3}{1 + K_1[A^-] + K_1K_2[A^-]^2 + K_1K_2K_3[A^-]^3} \tag{12}$$

From an experimental standpoint, $\bar{n}$ may be expressed in terms of the total glycine concentration (A_{tot}), the concentration of HA and A^-, and the total Ni^{2+} concentration (M_{tot}):

$$\bar{n} = \frac{A_{tot} - [HA] - [A^-]}{M_{tot}} \tag{13}$$

To determine $[HA]$ and $[A^-]$ in equation (13), an expression for the H^+ bound to the glycinate ion is introduced:

$$\text{Bound } H^+ = [HA] = \left(\begin{array}{c} \text{added } H^+ \\ \text{from } HNO_3 \end{array} \right) + \left(\begin{array}{c} H^+ \text{ from} \\ \text{dissociation} \\ \text{of } H_2O \end{array} \right) - \text{free } H^+$$

$$[HA] = C_H + [OH^-] - [H^+] \tag{14}$$

---

*J. Bjerrum, *Metal Ammine Formation in Aqueous Solution*, P. Haase and Son, Copenhagen, 1941, 1957.

where C_H is the concentration of HNO_3 added to the Ni^{2+} solution. Substitution of equation (7) gives

$$[A^-] = \frac{K_a}{[H^+]} (C_H + [OH^-] - [H^+]) \qquad (15)$$

Substitution of equations (14) and (15) into (13) yields:

$$\bar{n} = \frac{A_{tot} - \left(1 + \frac{K_a}{[H^+]}\right)(C_H + [OH^-] - [H^+])}{M_{tot}} \qquad (16)$$

Thus, $[A^-]$ and $\bar{n}$ may be calculated from experimentally known quantities.

In Figure 13–1 is shown a typical plot of the average number of ligands per metal ion, $\bar{n}$, *versus* the concentration of the free ligand, A^-. From such plots it is possible to estimate the values of K_1, K_2, and K_3. From equation (4), for example, it is noted that when $[NiA^+] = [Ni^{2+}]$, then $K_1 = 1/[A^-]$. Since the condition $[NiA^+] = [Ni^{2+}]$ means that the average number of ligands per metal ion is ½, i.e., $\bar{n} = ½$, the value of $[A^-]$ at $\bar{n} = ½$ from a plot as in Figure 13–1 will allow K_1 to be estimated. Likewise at $\bar{n} = 1½$ and $2½$, the values of $[A^-]$ will allow the estimation of K_2 and K_3. In general, K_n can be estimated from the expression: $K_n = \left[\frac{1}{[A^-]}\right]_{\bar{n}=n-\frac{1}{2}}$. It should be noted that these K_n values are only estimates because more than two complexes, e.g., more than just NiA^+ and Ni^{2+}, are generally present in a solution where $\bar{n} = n-½$.

A more precise graphical method of evaluating K_1, K_2, and K_3 from $\bar{n}$ and $[A^-]$ data has been developed by Rossotti and Rossotti* and

---

*F. J. C. Rossotti and H. S. Rossotti, *Acta Chem. Scand.*, 9, 1166 (1955).

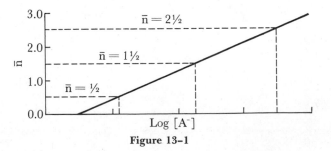

Figure 13–1

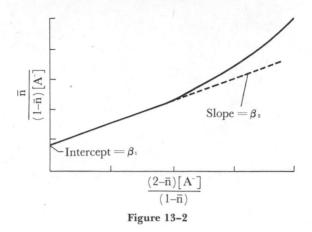

Figure 13–2

will be used in this experiment. First equation (12) is converted to equation (17) by introducing expressions for the overall stability constants, β:

$$\bar{n} = \frac{\beta_1[A^-] + 2\beta_2[A^-]^2 + 3\beta_3[A^-]^3}{1 + \beta_1[A^-] + \beta_2[A^-]^2 + \beta_3[A^-]^3} \qquad (17)$$

where $\beta_1 = K_1$, $\beta_2 = K_1K_2$, and $\beta_3 = K_1K_2K_3$. Rearrangement of this equation gives

$$\frac{\bar{n}}{(1-\bar{n})[A^-]} = \beta_1 + \frac{(2-\bar{n})[A^-]}{(1-\bar{n})}\beta_2 + \frac{(3-\bar{n})}{(1-\bar{n})}[A^-]^2\beta_3 \qquad (18)$$

Thus, a plot of $\dfrac{\bar{n}}{(1-\bar{n})[A^-]}$ *versus* $\dfrac{(2-\bar{n})[A^-]}{(1-\bar{n})}$ tends to a straight line at low $[A^-]$ of intercept β_1 and slope β_2 (Figure 13–2).

Using the value of β_1 obtained as the intercept, equation (18) is divided by $\dfrac{(2-\bar{n})}{(1-\bar{n})}[A^-]$ and rearranged to give

$$\frac{\bar{n} - (1-\bar{n})\beta_1[A^-]}{(2-\bar{n})[A^-]^2} = \beta_2 + \frac{(3-\bar{n})}{(2-\bar{n})}[A^-]\beta_3 \qquad (19)$$

The intercept of a plot of $\dfrac{\bar{n} - (1-\bar{n})\beta_1[A^-]}{(2-\bar{n})[A^-]^2}$ *versus* $\dfrac{3-\bar{n}}{2-\bar{n}}[A^-]$ gives β_2 and the slope is β_3 (Figure 13–3).

From these relatively precise values of β_1, β_2, and β_3, the stability constants K_1, K_2, and K_3 may be calculated. Because of the method of data treatment, the precisions of the constants decrease in the order $K_1 > K_2 > K_3$. Calculations of the functions required for the evaluation of the stability constants are, as you will experience, *very* time-consuming. When large numbers of such determinations are to be conducted, the calculations are normally computerized.

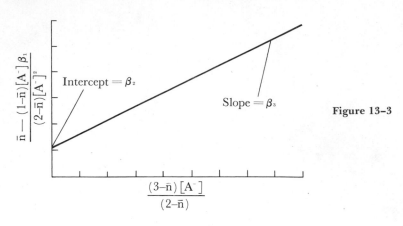

Figure 13–3

EXPERIMENTAL PROCEDURE

A pH meter is a sensitive instrument that must be operated properly and with care. The electrodes are particularly sensitive to physical breakage or cracking. When not in use, they should be stored in a buffer solution of approximately pH 4. Just prior to use they should be rinsed with distilled water (use a wash bottle) and blotted dry with a piece of filter paper. They may then be introduced into the solution under study.

The pH meter must first be standardized. Insert the electrodes into a pH 7 buffer and adjust the calibration knob until the meter reads the same as the pH of the buffer. Then, using buffers of pH 4 and 10, check the meter reading. If the reading differs by more than 0.1 pH unit from that of the buffer, recheck the meter reading of the pH 7 buffer. If the adjustment for pH is correct at pH 7 yet the reading is incorrect at pH 4 and 10, the meter reading will have to be corrected for this inaccuracy. This may be done on some pH meters by adjusting the slope adjustment or the temperature compensator. Otherwise, it should be accomplished by determining a meter correction constant, C_{meter}, from the linear expression:

$$pH = 7.00 + (pH_{meter} - 7.00)C_{meter} \qquad \textbf{(20)}$$

where pH is the correct pH, and pH_{meter} is the meter reading. Using the pH 4 or 10 buffer, determine C_{meter}. To obtain correct pH readings in all subsequent measurements, pH_{meter} readings will have to be corrected with equation (20).

Prepare the following solutions, using distilled water:

  20 ml 0.4 M glycine, HA
500 ml 0.2 M KNO_3
100 ml 0.10 M HNO_3
100 ml 0.50 M NaOH

Standardize the NaOH solution by titrating a solution of the primary standard, potassium acid phthalate, using phenolphthalein indicator. Use the standardized NaOH solution to standardize the HNO_3 solution. All solutions (particularly NaOH) should be stored in tightly stoppered bottles to prevent atmospheric CO_2 from dissolving to give H_2CO_3, H^+, HCO_3^-, and CO_3^{2-}.

First the pK_a of glycine, HA, must be determined. Add a magnetic stirring bar and introduce the electrodes of the standardized pH meter into a 400 ml beaker containing 90 ml of H_2O, 100 ml of 0.2 M KNO_3, and 10 ml of 0.4 M glycine. If possible, it is desirable to thermostat the beaker at a constant temperature with a small water bath or jacketed beaker. However, the K values have a relatively small temperature dependence, and even a complete lack of thermostatting will introduce only small errors. Rinse and fill a 10 ml burette with the standardized 0.5 M NaOH. With stirring, add about 15 aliquots of NaOH in titrating from 0 to 100 per cent of the glycine. After each addition, record the amount of NaOH solution added and the pH. Using the points in the 20 to 80 per cent titration range, calculate the K_a of glycine from equations (20), (9), (10), and (11). It should be noted that in the near neutral solutions, $[H^+]$ and $[OH^-]$ are very small compared to $[Na^+]$ and may be neglected in the last term of equation (11). Obtain an average K_a value.

Next, the data necessary for calculating K_1, K_2, and K_3 will be collected. Prepare 20 ml of 0.4 M sodium glycinate solution by exactly neutralizing weighed solid glycine with a calculated amount of standardized 0.5 M NaOH and diluting with water to a total volume of 20 ml. Rinse and fill the cleaned 10 ml burette with this solution. Using this sodium glycinate solution, titrate, with stirring, a solution of 100 ml of 0.2 M KNO_3, 1 millimole of $NiCl_2 \cdot 6H_2O$ (a good grade of this compound may be weighed out directly without standardization), 10 ml of 0.10 M HNO_3, and 90 ml of H_2O in a 400 ml beaker. After each 0.2 ml aliquot addition, record the amount of glycinate solution added and the pH of the solution. Add the full 10 ml. Since the volume (10 ml at the end of the titration) of sodium glycinate solution added is small compared to the volume (200 ml) of the Ni^{2+} solution, correcting the concentrations for dilution is not necessary. From the titration data, calculate $[A^-]$ from equation (15) and $\bar{n}$ from equation (16) at each of the 50 points. Note that $[H^+]$ and $[OH^-]$ may be negligible in some cases. In these calculations, be sure to correct the meter reading (equation (20)), if necessary, and to convert pH to $[H^+]$ using equations (9) and (10). Make a plot of $\bar{n}$ *versus* log $[A^-]$ as in Figure 13-1. From the values of $[A^-]$ at n = ½, 1½, and 2½, calculate approximate values of K_1, K_2, and K_3.

Now calculate $\dfrac{\bar{n}}{(1-\bar{n})[A^-]}$ and $\dfrac{(2-\bar{n})[A^-]}{(1-\bar{n})}$ for those points with values of $\bar{n}$ between 0.8 and 0.2. Plot these quantities as in Figure 13–2 to obtain an accurate value of β_1 from the intercept. Then evaluate $\dfrac{\bar{n} - (1 - \bar{n})\beta_1[A^-]}{(2 - \bar{n})[A^-]^2}$ and $\dfrac{(3 - \bar{n})[A^-]}{(2 - \bar{n})}$ for the points whose $\bar{n}$ values range from 1.1 to 1.7. From the plot of these quantities as in Figure 13–3, determine β_2 (intercept) and β_3 (slope). Finally, calculate K_1, K_2, and K_3 from β_1, β_2, and β_3, and compare these accurate values with the approximate ones that you evaluated earlier by using the half-$\bar{n}$ method.

REPORT

Include the following:
1. Plot of the titration curve, pH *versus* volume of NaOH, for glycine.
2. Average pK_a value of glycine in 0.10 M KNO_3.
3. Plot of $\bar{n}$ *versus* log $[A^-]$. Approximate K_1, K_2, and K_3 values from half-$\bar{n}$ method.
4. Plot used to determine β_1.
5. Plot used to determine β_2 and β_3.
6. Accurate values of K_1, K_2, and K_3 in 0.10 M KNO_3.
7. Comment on the relative precisions of the K_1, K_2, and K_3 values.

QUESTIONS

1. Draw all geometrical and optical isomers of the three complexes, Ni(glycinate)$^+$, Ni(glycinate)$_2$, and Ni(glycinate)$_3{}^-$.
2. Derive equation (11).
3. A solution prepared by mixing 100 ml of 0.2 M glycine and 100 ml of 0.2 M Cu^{2+} is adjusted to a pH of 7 with NaOH. Calculate approximate percentages of the original Cu^{2+} that exists as Cu^{2+}, CuA^+, and CuA_2 in the solution. Use the pK_a that you determined for glycine in this experiment and the following stability constants:

$$Cu^{2+} + A^- \rightleftarrows CuA^+ \qquad \log K_1 = 8.38$$

$$CuA^+ + A^- \rightleftarrows CuA_2 \qquad \log K_2 = 6.87$$

4. You wish to determine the formation constants (K_1, K_2, and K_3) for the coordination of ethylenediamine, $NH_2CH_2CH_2NH_2$ (en), to Ni^{2+} to form Ni(en)$^{2+}$, Ni(en)$_2{}^{2+}$, and Ni(en)$_3{}^{2+}$.
   (a) How would you determine the first and second pK_a values of en?
   (b) Define these two pK_a values. Which of the two will be larger, and why?
   (c) Briefly explain what titrations you would carry out in order to determine the stability constants K_1, K_2, and K_3. Would the procedure differ from that used in this experiment?
   (d) Which value would you expect to be larger, K_1 or K_2?

5. Consider the coordination of Cu^{2+} by the tridentate iminodiacetate ligand, $HN(CH_2CO_2^-)_2$, labeled A^{2-}. When the monoprotonated form, $H_2\overset{+}{N}(CH_2CO_2^-)_2$, labeled HA^-, is added to an aqueous solution of Cu^{2+}, the following equilibrium is rapidly established:

$$Cu^{2+} + HA^- \rightleftharpoons CuA + H^+$$

If a solution prepared by combining 100 ml of 0.02 M Cu^{2+} and 100 ml of 0.02 M HA^- is adjusted to pH 1.60 with NaOH, what are the concentrations of Cu^{2+}, HA^-, and CuA in the solution? (The K_a for HA^- is 4.70×10^{-10}, and K_1 for $Cu^{2+} + A^{2-} \rightleftharpoons CuA$ is 4.26×10^{10}.) Draw the probable structure of the CuA complex.

6. The pH method of measuring equilibrium constants is a bulk technique in that it does not discriminate between geometrical isomers. Suggest techniques that would permit separate detection of isomers.

INDEPENDENT STUDIES

A. Determine stability constants for the coordination of Ni^{2+} by another amino acid, such as alanine or valine.

B. Determine stability constants for the coordination of glycine at a higher ionic strength, such as 1.0 M KNO_3. Compare the results with those at 0.1 M KNO_3.

C. Determine stability constants for the coordination of Cu^{2+} (or other metal ions) by glycine.

D. Using the method of continuous variations (Job's method, see Experiment 12), establish the compositions of the Ni(glycinate)$_n^{(2-n)+}$ complexes present in solutions of Ni^{2+} and glycinate$^-$.

E. Determine stability constants for the stepwise coordination of Cu^{2+} by NH_3. (W. B. Guenther, *J. Chem. Educ.*, *44*, 46 (1967).)

F. Determine stability constants for the stepwise coordination of Cu^{2+} by ethylenediamine, $NH_2CH_2CH_2NH_2$. (D. E. Goldberg, *J. Chem. Educ.*, *39*, 328 (1962).)

G. Determine stability constants for the stepwise coordination of the uranyl ion, UO_2^{2+}, by acetate. (D. R. Williams, *J. Chem. Educ.*, *48*, 480 (1971).)

H. Determine pK_a values of the trifunctional amino acid, cysteine. (G. E. Clement and T. P. Hartz, *J. Chem. Educ.*, *48*, 395 (1971).)

REFERENCES

Ni(glycinate)$_n^{(2-n)+}$

N. C. Li, J. M. White, and R. L. Yoest, *J. Amer. Chem. Soc.*, *78*, 5218 (1956).

Stability Constant Determination Techniques and Data Treatment

A. Albert and E. P. Serjeant, *The Determination of Ionization Constants,* Chapman and Hall Ltd., London, 1971. A laboratory manual of methods.

M. T. Beck, *Chemistry of Complex Equilibria,* Van Nostrand Reinhold, New York, 1970. Experimental and computational methods of evaluating stability constants.

A. M. Bond, *Coord. Chem. Rev., 6,* 377 (1971). A comparison of different calculation procedures for evaluating stability constants.

R. F. Cookson, *Chem. Rev., 74,* 5 (1974). Methods of determining acid dissociation constants.

D. R. Crow, *Polarography of Metal Complexes,* Academic Press, New York, 1969. Includes polarographic methods of determining stability constants.

S. Froneaus in *Technique of Inorganic Chemistry, Vol. I,* Interscience Publishers, New York, 1963, pp. 1–36. Methods of determining stability constants of complexes.

L. Johansson, *Coord. Chem. Rev., 3,* 293 (1968). Use of solubility measurements to determine stability constants.

M. M. Jones, *Elementary Coordination Chemistry,* Prentice-Hall, Englewood Cliffs, New Jersey, 1964, Chapter 8.

F. J. C. Rossotti and H. Rossotti, *The Determination of Stability Constants,* McGraw-Hill, New York, 1961. Standard reference to methods of data treatment.

Stability Constants of Complexes

S. Chaberek and A. E. Martell, *Sequestering Agents,* Wiley-Interscience, New York, 1959. Applications of complexing ligands.

J. F. Hinton and E. S. Amis, *Chem. Rev., 71,* 627 (1971); S. F. Lincoln, *Coord. Chem. Rev., 6,* 309 (1971). Solvent coordination numbers of metal ions in solution.

G. Schwarzenbach in *Advances in Inorganic Chemistry and Radiochemistry, Vol. 3,* Academic Press, New York, 1961, pp. 257–286. Discussion of trends in stability constants.

L. G. Sillen and A. E. Martell, eds., *Stability Constants of Metal-Ion Complexes,* Chemical Society of London, 1964. A thorough compilation of stability constants of metal ions with inorganic and organic ligands.

$[1,3,5-C_6H_3(CH_3)_3]Mo(CO)_3$ and $(C_6H_6)Cr(CO)_3$

Although it had been known for years that the π-bond in olefins was essential for their coordination to transition metals, it was not until approximately 1955 that benzene was also found to form stable complexes. Assuming a Kekulé structure for benzene, one might expect benzene to act like three olefins and form three donor bonds with the metal. As illustrated by the present experiment, benzene and its derivatives do often react to displace three donor ligands. The three ligands that will be displaced in this synthesis are CO groups. Because mesitylene, $1,3,5-C_6H_3(CH_3)_3$, forms more stable complexes than benzene, it will be used first as the aromatic ligand.

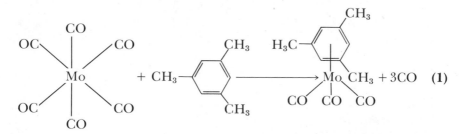

The reaction may be carried out simply by refluxing $Mo(CO)_6$ in mesitylene. The product, $[1,3,5-C_6H_3(CH_3)_3]Mo(CO)_3$, has a sandwich structure in which the benzene ring lies parallel to the plane of the three carbon atoms of the CO groups, as shown in equation (1). It may be considered to be an octahedral complex since all of the OC–Mo–CO bond angles are very nearly 90° as they are in $Mo(CO)_6$. The arene group sits on one face of the octahedron. While our earlier comments suggested that benzene acts like three olefinic ligands, x-ray structural studies show that all of the C–C bond distances in the benzene ring are equal, and there is no strong evidence for localized double bonds. A molecular orbital bonding scheme involving molecular orbitals of the aromatic benzene and s, p, and d orbitals on the metal qualitatively accounts for the structure and spectral properties of these compounds.

129

The second and somewhat more challenging synthesis is that of benzene tricarbonylchromium(0), $(C_6H_6)Cr(CO)_3$:

$$Cr(CO)_6 + C_6H_6 \rightarrow (C_6H_6)Cr(CO)_3 + 3CO \qquad (2)$$

The difference between this synthesis and that of the mesitylene complex is the lower boiling point of benzene (80°C). Since its reaction with $Cr(CO)_6$ requires temperatures of 160 to 175°C, benzene will normally distill away from the reaction mixture before any reaction occurs. Thus, the reaction must be conducted in a bomb in order to prevent the benzene from evaporating. In the bomb, the high reaction temperature will vaporize some of the benzene, causing an increase in pressure, but most of the benzene will remain liquid and available for reaction with the $Cr(CO)_6$. The bomb technique is used in a variety of organometallic syntheses.

By reaction of $Mo(CO)_6$ or $Cr(CO)_6$ with other aromatic molecules, it is possible to synthesize compounds with novel geometries:

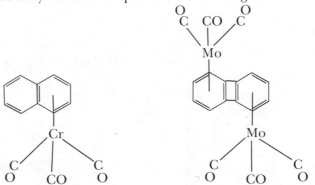

These and many other structures have been established by x-ray diffraction studies. The analogy of borazole, $B_3N_3H_6$ (sometimes called "inorganic benzene"), to benzene has resulted in the recent synthesis of the hexamethylborazole analog of $[C_6(CH_3)_6]Cr(CO)_3$:

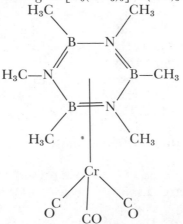

Of some interest to organometallic chemists is the aromaticity of benzene in these complexes compared with that of the free molecule. A chemical indication of the aromaticity of a benzene derivative is its rate of acetylation under Friedel-Crafts reaction conditions. Since this is mechanistically an electrophilic substitution reaction involving CH_3CO^+, reaction occurs faster with benzenes having a higher electron density in the π-system. The reaction with $(C_6H_6)Cr(CO)_3$

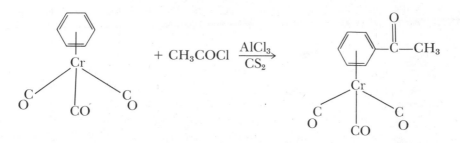

occurs readily but at a rate that is slower than for benzene itself. This suggests that the electron density in the benzene is depleted by the $Cr(CO)_3$ moiety upon complexation. This is expected if benzene, like most ligands, acts as an electron donor toward transition metals.

While the complexes, (benzene)$M(CO)_3$ where M = Cr, Mo, or W, have been studied most extensively because of their relatively high stabilities, the *bis*-benzene complexes of these metals were the first known complexes of benzene. The standard synthesis of $Cr(C_6H_6)_2$ is carried out according to the following equations:

$$3CrCl_3 + 2Al + AlCl_3 + 6C_6H_6 \rightarrow 3Cr(C_6H_6)_2{}^+AlCl_4{}^-$$

$$2Cr(C_6H_6)_2{}^+ + S_2O_4{}^{2-} + 4OH^- \rightarrow Cr(C_6H_6)_2 + 2SO_3{}^{2-} + 2H_2O$$

In the first reaction, the $AlCl_3$ acts to remove Cl^- from the Cr as $AlCl_4{}^-$, and powdered Al reduces the Cr(III) to Cr(I). The cationic complex, $Cr(C_6H_6)_2{}^+$, is a stable material that has the ferrocene sandwich structure. It may be reduced with dithionite ion, $S_2O_4{}^{2-}$, to give $Cr(C_6H_6)_2$ in an overall yield of 60 per cent. *bis*-Benzenechromium(0) has the expected sandwich structure:

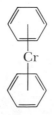

It may be readily oxidized by O_2 to the cation, $Cr(C_6H_6)_2^+$. Using other transition metal halides, it is possible to prepare a variety of other *bis*-benzene complexes by reaction with Al and $AlCl_3$ followed by reduction with $S_2O_4^{2-}$. Some of these are $V(C_6H_6)_2^+$, $V(C_6H_6)_2$, $Mo(C_6H_6)_2^+$, $Mo(C_6H_6)_2$, $W(C_6H_6)_2^+$, $W(C_6H_6)_2$, $Re(C_6H_6)_2^+$, $Fe(C_6H_6)_2^{2+}$, $Ru(C_6H_6)_2^{2+}$, and $Co(C_6H_6)_2^+$. The variety and novelty of arene complexes have made this area of research one of the most interesting in inorganic chemistry. For their contributions to the general area of organometallic chemistry, E. O. Fischer and G. Wilkinson were awarded the Nobel Prize in 1973.

EXPERIMENTAL PROCEDURE

Metal carbonyls, in general, are toxic compounds and should be handled with care. They are particularly dangerous because of their relatively high volatilities. For this reason, $Ni(CO)_4$ (b.p., 43°C) is exceedingly hazardous. On the other hand, solid carbonyl complexes such as $Cr(CO)_6$, $Mo(CO)_6$, and $W(CO)_6$ do not present as serious a danger, but they should be handled in a *hood*. In reactions (1) and (2), carbon monoxide gas is evolved. The volume of CO given off is relatively small, however, and if the reactions are conducted in a *hood,* the CO will not be present in sufficiently high concentrations to be dangerous.

Part A. $[1,3,5-C_6H_3(CH_3)_3]Mo(CO)_3$, Mesitylene Tricarbonylmolybdenum(0)

The reaction will be conducted in the apparatus shown in Figure 14–1. First put 2.0 g (7.6 mmoles) of $Mo(CO)_6$ and 10 ml (72 mmoles) of mesitylene (b.p., 165°C) in the 50 ml round-bottom flask with a side-arm and stopcock. The apparatus is assembled in the *hood* as in Figure 14–1, using a straight reflux condenser of approximately 30 cm length. The $Mo(CO)_6$ is volatile and will sublime into the condenser during the reaction. For this reason it is desirable for the mesitylene vapors to rise high into the condenser to return any sublimed $Mo(CO)_6$ to the reaction flask. To allow the mesitylene to wash the $Mo(CO)_6$ into the flask, do not cool the condenser with water. The temperature of the room will provide adequate cooling.

The tendency of $Mo(CO)_6$ and the product to react with oxygen at high temperatures requires that the reaction be conducted in an inert atmosphere. Connect the side-arm to a nitrogen cylinder *via* a length of rubber tubing. Flush the apparatus with a moderate stream of nitrogen for approximately 5 minutes. Turn off the nitrogen and then close the stopcock. Heat the solution at a moderate boil for about 30 minutes with a rheostat-controlled heating mantle. Then remove the

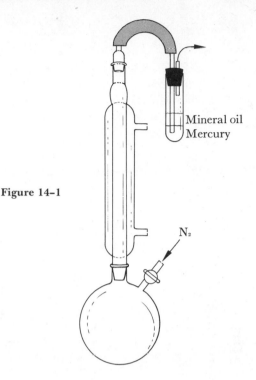

Mineral oil
Mercury

Figure 14–1

N$_2$

heating mantle and immediatcly turn on the nitrogen flush to prevent the mercury and mineral oil from being drawn back into the reaction mixture. The nitrogen flush also serves to sublime out of solution any remaining Mo(CO)$_6$.

When the solution has cooled to room temperature, turn off the nitrogen stream and dismantle the apparatus. Add 15 ml of hexane (or hydrocarbon fraction of 60 to 70°C boiling range) to complete the pre-cipitation. Suction filter (Figure 1–1) the solution on a medium frit and rinse, with 5 ml of hexane, the yellow product that is contaminated with black metallic molybdenum. Purify the crude product by dissolving it in a minimum of CH$_2$Cl$_2$ ($\sim$ 10 ml). After filtering the solution, add 25 ml of hexane to precipitate the product. Suction filter off the yellow [1,3,5–C$_6$H$_3$(CH$_3$)$_3$]Mo(CO)$_3$, wash twice with 4 ml of hexane, and allow the product to dry on the frit while under aspiration. (A second batch of product can be obtained by reducing the volume of the mother liquor under water aspirator vacuum [see Figure 6–3] at room temperature.) Calculate the percentage yield. If desired, the product may be further purified by sublimation at approximately 120°C, under high vacuum (Figure 11–3). Since [1,3,5–C$_6$H$_3$(CH$_3$)$_3$]Mo(CO)$_3$ will decompose over a period of weeks in the presence of light and air, it should be stored in the dark in a tightly stoppered container that has been flushed with nitrogen.

Part B. (C₆H₆)Cr(CO)₃, Benzene Tricarbonylchromium(0)

A high pressure bomb (also called an autoclave) is frequently used to carry out reactions of gases at high pressures so that the concentration of the gas is much higher than it would be at one atmosphere pressure. It is also used in reactions in which one or more of the reactants would be a gas at atmospheric pressure but is a liquid in a bomb under pressure. In the present experiment, the bomb is used for the latter purpose, namely, to maintain the reactant benzene as a liquid in contact with $Cr(CO)_6$ even at the high reaction temperatures (160 to 175°C).

Just the name *high pressure bomb* is sufficient to suggest that this is a dangerous experiment. In fact, there is no hazard associated with this experiment if the instructions are followed carefully. The pressure developed in the bomb in this experiment is due to the vapor pressure of benzene at the reaction temperature and to the gaseous CO released during the reaction (equation (2)). The maximum pressure developed in the bomb in this experiment is only 250 to 400 psi (pounds per square inch), far below the maximum recommended pressures, generally 2000 to 3000 psi, of such bombs. This is a safe introduction to the use of high pressure bombs.

One type of bomb that may be used in this experiment is shown in Figure 14–2. It is a simple split-ring closure, general purpose bomb that consists of a cylinder and a head, to which is attached an inlet-outlet valve and pressure gauge. The pressure gauge unit also contains a safety blowout (rupture) disc in case the pressure exceeds the recommended maximum pressure. See the references at the end of this chapter for additional information on high pressure bombs. In the Notes to Instructor section at the end of this book is given further information on the bomb shown in Figure 14–2. Regardless of what specific bomb is used, read the manufacturer's instructions carefully and understand how to assemble and use it. Also know its pressure limits.

The reaction mixture will not be placed into the bomb directly but into a glass liner instead. The liner is simply a glass test tube that fits snugly into the bomb, leaving about 2 cm between the top of the liner and the bomb head.

Place about 2.0 g of $Cr(CO)_6$ in the glass liner. Saturate about 40 ml of benzene with N_2 by bubbling the gas through it for several minutes. Pour about 10 ml of the benzene into the bomb. Then insert the glass liner into the bomb. This will displace the benzene from the bomb, and it will overflow the top of the liner. This ensures that the space between the liner and the bomb is filled with benzene. (When the closed bomb is heated, the benzene will vaporize, cool on the unheated top part of the bomb, and condense. Some of the condensed benzene will run down the sides into the space between the liner and the bomb if benzene is not already there. Having benzene in this space avoids the possibility of all of the benzene distilling out of the reaction liner.)

Figure 14–2   (Courtesy of Parr Instrument Co.)

Add the remainder of the N_2-saturated benzene to the glass liner and flush the container well with N_2. Place the head on the bomb, making certain that the gasket is seated properly. Put on the split rings and collar, and secure the top bolts by alternately tightening opposite bolts in the head. (See the manufacturer's instructions.) Evacuate the bomb by attaching it to a water aspirator for 1 or 2 minutes (not longer). Then close the bomb valve and disconnect the aspirator. The resulting partial vacuum in the bomb will reduce the pressure build-up that occurs during the reaction.

Lower the bomb, in a vertical position as in Figure 14–2, into a silicone fluid oil bath to a depth that is above the level of the reactants in the bomb. Heat the bomb in the bath at 160 to 175°C for 24 to 48 hours.

Place a safety screen around the bomb. It is also a good idea to attach a sign warning others to leave the reaction undisturbed. After the reaction period, remove the bomb and allow it to cool to room temperature (~2 hours); release the pressure by slowly opening the bomb valve in the hood(!). (The amount of CO produced in the reaction is small, and no hazard is involved in venting the bomb in a properly functioning hood.)

Then remove the bomb head and glass liner. Pour the benzene solutions in the liner and the bomb into a flask (such as a suction flask or that shown in Figure 6–3). Evaporate the solution to dryness under a water aspirator vacuum. The yellow solid that remains is a mixture of $(C_6H_6)Cr(CO)_3$ and $Cr(CO)_6$. Take an infrared spectrum of the mixture in CS_2 solvent.

Purify the $(C_6H_6)Cr(CO)_3$ by dissolving it in a minimum volume of boiling hexane; filter the hot solution and cool it in an ice bath to precipitate the product. Measure the infrared spectrum of the yellow product.

The product may also be purified by column chromatography on silica gel (see Experiment 16), or by subliming the $Cr(CO)_6$ at ~40°C under vacuum and then raising the oil bath temperature to 70 to 90° to obtain the sublimed $(C_6H_6)Cr(CO)_3$ (see Figure 11–3).

Infrared Spectra of Organometallic Complexes in Solution

Determine the infrared spectra of the compounds by dissolving a few small crystals in approximately 2 ml of CS_2. With an eye-dropper, introduce the solution into an infrared cell such as that in Figure 14–3. (A discussion of infrared spectroscopy may be found in Experiment 1, p. 19.) Unless the cell is sealed well, volatile solvents such as CS_2 will rapidly leak from the cell. In this case, $CHCl_3$ solvent may be used, but the C—H stretching region will be obscured by solvent absorption. Place the sample cell and a matching cell containing only solvent in the proper light beams of the spectrophotometer and record the spectrum. If the sample and reference cells are not well matched, solvent absorptions may appear in the spectrum. Their presence may be checked by measur-

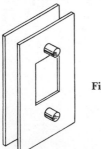

Figure 14–3

ing the spectrum obtained when both cells contain only solvent. After taking the spectrum, pour the solution out of the cell. Clean it by filling with solvent and decanting. The washing procedure should be repeated at least one more time. After drying the cells with a rapid flush of dry air or nitrogen, return them to the desiccator. If sample solution is spilled on the outside surface of the cell window, rinse it off immediately with solvent. Otherwise, subsequent samples measured in that cell will always give a spectrum of your compound.

The IR spectra are dominated by the characteristically intense C—O stretching absorptions in the region near 2000 cm⁻¹. Assign the remaining absorption bands to the mesitylene and benzene portions of the molecules.

Proton Nuclear Magnetic Resonance Spectra of Organometallic Complexes

Electrons behave as if they were spinning spheres of negative charge. Experimental measurements suggest that electrons may spin in clockwise or counter-clockwise directions around an axis to generate a magnetic field, ↑ or ↓. The energy of an electron is the same regardless of whether it has a ↑ or ↓ spin. In the same way, many nuclei, e.g., ¹H, ³¹P, and ¹⁹F, spin in either of two ways to generate ↑ or ↓ magnetic fields. While nuclei of opposite spins normally have the same energies, in an externally applied magnetic field this is no longer true. Then the field produced by the spinning nucleus is either aligned with the applied field or opposes it. The nucleus aligned (↑) with the applied field is of lower energy than that opposing (↓) it.

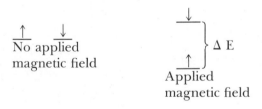

These spin states are separated by a relatively low energy ΔE. By irradiating a nucleus with radio frequency light, it is possible to cause the nucleus to go from the low energy state (↑) to that of higher energy (↓) with concomitant absorption of a radio frequency having the energy ΔE. The magnitude of ΔE depends upon the strength of the applied magnetic field, the spin properties of the nucleus under study, and the extent of shielding of the spinning nucleus (↑) from the applied magnetic field by electrons surrounding the nucleus.

As noted in the experiment on magnetic susceptibility, electrons will move in their orbitals so as to create a magnetic field that opposes any applied field. This electron density, which is determined by the chemical environment of the nucleus in a molecule, is a major factor influencing the value of ΔE. For this reason, chemically different ^1H atoms in a molecule will have different transition energies ΔE. For example, in ethanol, CH_3CH_2OH, there are three types of ^1H nuclei because of the three chemically different types of hydrogen atoms. The $\uparrow$ to $\downarrow$ transition will occur at different ΔE values for each kind of ^1H. Thus, an nmr spectrum of ethanol will show absorption of three energies of radio frequency radiation. Moreover, the amount of absorbed radiation depends upon the number of each kind of H atom. For CH_3CH_2OH, the ratio of the amount of absorbed light for the CH_3 protons, the CH_2 protons, and the OH proton is 3:2:1. This allows one to literally count the numbers of hydrogen atoms that are chemically different in a molecule.

In practice, it is more convenient to design nuclear magnetic resonance instruments in which the radio frequency (usually 60 megacycles per second) is held constant and the magnetic field is varied. An increase or decrease in the applied magnetic field strength produces a corresponding increase or decrease in ΔE. Thus, ΔE is varied until it is the same as the energy of the radio frequency radiation. Then absorption of the radiation occurs. In general, protons that are highly shielded by electrons require high magnetic fields for absorption to occur. The low resolution nmr spectrum of ethanol is shown in Figure 14–4.

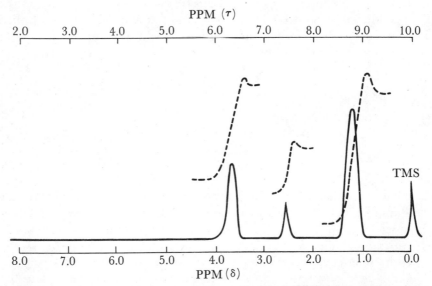

Figure 14–4 Low resolution nmr spectrum (——) of CH_3CH_2OH and integration (----).

The quantitative interdependence of the radio frequency and the applied magnetic field means that ΔE may be expressed as a radio frequency at a fixed magnetic field or in gauss at a fixed frequency. For purposes of specifying ΔE, it is convenient to do this as a radio frequency, cycles per second (Hertz, Hz, is commonly used for cps). Rather than simply specifying the absolute frequency, ν_s, required for the $\uparrow$ to $\downarrow$ transition for a given proton, it is usually expressed in terms of a difference between ν_s and ν_r of some reference compound, $(\nu_s - \nu_r)$. This difference divided by the frequency of the instrument (usually 60,000,000 Hz) is called the *chemical shift, δ*:

$$\delta = \frac{\nu_s - \nu_r}{60 \times 10^6} \qquad (3)$$

The reference compound that is frequently used for ^{1}H nmr spectroscopy is tetramethylsilane (TMS), Si(CH$_3$)$_4$. Equation (3) thus fixes the chemical shift of TMS at 0. Since the difference, $(\nu_s - \nu_r)$, usually ranges between 0 and -500 Hz, δ values normally extend from 0 to -8×10^{-6}. To avoid using the 10^{-6} factor, δ's are expressed as parts per million, e.g., -8.0 ppm (Figure 14–4). Chemical shifts are also frequently reported in τ units rather than δ units (Figure 14–4). The τ scale is the same as that for δ except that TMS is now assigned a τ value of 10 ppm. These scales are simply related by the equation:

$$\tau = 10 + \delta$$

The τ scale has the advantage that almost all chemical shifts of ^{1}H nuclei in known molecules have positive τ values.

The purpose of this exceedingly brief treatment of nmr has been to relate the parameters obtained from an nmr spectrum to the origins of the nuclear magnetic resonance phenomenon. The discussion has greatly oversimplified the theoretical background upon which nmr spectroscopy is based. It has also entirely ignored the vast area of spin-spin coupling, which is of extreme interest and importance. Moreover, we have discussed only one isotope, ^{1}H; extensive nmr studies of ^{11}B, ^{31}P, and ^{19}F nuclei in a wide range of inorganic compounds have contributed significantly to our understanding of structure and bonding. (Isotopes of other nuclei having nuclear spins are given in Appendix 7.) The reader is very much encouraged to examine the references listed at the end of this experiment.

Almost all commercially available nmr instruments require liquid samples of approximately 1 ml volume. If the compound is a liquid, its spectrum can be measured neat. Since many samples are solid, they must first be dissolved in a solvent that itself does not absorb in the region of the nmr spectrum of interest. Generally, these solutions must be relatively concentrated, approximately 0.1 to 0.2 g of compound in 1 ml of solvent. A drop of TMS is added to the solution for positioning of $\delta = 0$ ppm.

Make as concentrated a $CHCl_3$ solution of $[1,3,5\text{-}C_6H_3(CH_3)_3]\text{-}Mo(CO)_3$ as possible. (Note that commercial $CHCl_3$ contains about 1 per cent ethanol as stabilizer.) Measure its nmr spectrum immediately, because some decomposition will occur on standing. While it is very desirable for the spectrum of the student's sample to be determined, the unavailability of an nmr instrument may make this unfeasible. For this reason, the nmr spectrum that we have determined for this compound has been included (Figure A) in the section on NMR and Mass Spectra near the end of the book. The student should interpret this spectrum.

Mass Spectra of Organometallic Complexes

A schematic diagram of a mass spectrometer is shown in Figure 14–5. The entire system is under high vacuum. The sample must either be a gas or be sufficiently volatile to be converted to a gas at elevated temperatures. The high sensitivity of the instrument allows samples with exceedingly low volatilities to be examined. Ionic compounds and high polymers are two of the few types of materials that are very difficult to study in a mass spectrometer.

The sample is introduced into the inlet system and diffuses into the ionization chamber, where it is bombarded by the beam of electrons from the electron gun. If the electrons are sufficiently energetic, they may simply ionize the molecule, M:

$$M + e^- \rightarrow M^+ + 2e^-$$

$$\text{or } M + e^- \rightarrow M^{2+} + 3e^-$$

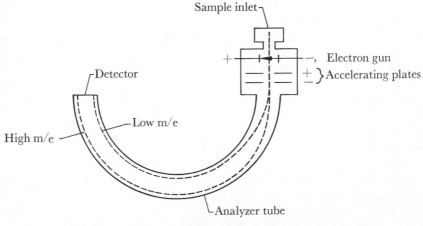

Figure 14–5

The resulting ion will diffuse to the accelerating plates, where it will be given a known kinetic energy by the known voltage between these plates. The accelerated ions then pass into the analyzer tube. Perpendicular to this tube are mounted the pole pieces of a magnet. Since a magnetic field bends the path of a moving charged particle, the ions follow curved paths, as shown in Figure 14–5.

The radius of curvature is determined by the mass (m) and the charge (e) of the ion. Light-weight ions and those with high charge will have the smallest radius of curvature. In other words, the radius of curvature will be smallest for ions of low m/e and greatest for those of high m/e. For a group of +1 ions, the radius of curvature and the position at which they strike the detector will only depend upon the mass of M^+. The accuracy of mass measurements by this technique is determined by the nature of the instrument, but in most cases masses of ± 0.5 atomic mass unit (a.m.u.) are easily discriminated.

Consider the mass spectrum of Ni(CO)$_4$. The initial ionization process produces Ni(CO)$_4^+$ ions that are detected. Nickel exists as several stable isotopes with natural abundances as follows:

Isotope	Natural Abundance
^{58}Ni	67.77 per cent
^{60}Ni	26.16 per cent
^{61}Ni	1.25 per cent
^{62}Ni	3.66 per cent
^{64}Ni	1.16 per cent

The total mass of the Ni(CO)$_4$ molecule containing the ^{58}Ni isotope is 170, assuming the remainder of the molecule to contain only ^{12}C and ^{16}O isotopes. Since Ni(CO)$_4^+$ containing the other Ni isotopes will also appear in the spectrum, ions of mass 170, 172, 173, 174, and 176 will be present, and their relative concentrations as measured by the peak heights in the spectrum should be proportional to the relative abundances of the Ni isotopes. Thus, in the line diagram spectrum shown in Figure 14–6, the most intense peak of the parent ion (Ni(CO)$_4^+$) occurs at a mass of 170 and the next most intense peak is that of mass 172.

To this point, it has been assumed that only ^{12}C and ^{16}O isotopes are present. For O, this is essentially true since 99.76 per cent of naturally occurring oxygen is ^{16}O. For C, this is not quite so true.

Isotope	Natural Abundance
^{12}C	98.89 per cent
^{13}C	1.11 per cent

The measurable abundance of ^{13}C means that measurable amounts of ^{58}Ni(^{12}CO)$_3$(^{13}CO)$^+$ (mass 171) should be visible in the mass spectrum.

Since 1.1 per cent of the C is ^{13}C, there is a 4.4 per cent probability that one of the four C atoms in $Ni(CO)_4$ is ^{13}C. In other words, 4.4 per cent of $^{58}Ni(CO)_4$ is $^{58}Ni(^{12}CO)_3(^{13}CO)$. Because the probability of finding two ^{13}CO groups in the same $^{58}Ni(CO)_4$ molecule is so very small, the percentage of $^{58}Ni(^{12}CO)_2(^{13}CO)_2$ is only 0.07 per cent, a negligible amount in most spectra.

As is obvious from Figure 14-6, a mass spectrum contains more than just parent ion peaks. If the bombarding electrons in the spectrometer are of sufficiently high energy, fragmentation of the molecule also occurs. For example,

$$Ni(CO)_4 + e^- \rightarrow Ni(CO)_4^+ + 2e^-$$

$$Ni(CO)_4 + e^- \rightarrow Ni(CO)_3^+ + CO + 2e^-$$

$$Ni(CO)_4 + e^- \rightarrow Ni(CO)_2^+ + 2CO + 2e^-$$

$$Ni(CO)_4 + e^- \rightarrow Ni(CO)^+ + 3CO + 2e^-$$

$$Ni(CO)_4 + e^- \rightarrow Ni^+ + 4CO + 2e^-$$

and also

$$Ni(CO)_4 + e^- \rightarrow Ni(CO)_3 + CO^+ + 2e^-$$

All of the ions containing Ni, with their characteristic isotopic mass distributions, are found in the spectrum.

In addition to the $+1$ ions, $+2$ ions are sometimes observed, but they are generally of low abundance. For example, the $+2$ ion, $^{58}Ni(^{12}CO)_4^{2+}$, should occur at an m/e value of $170/+2 = 85$ a.m.u. The spectrum shows no peak at that mass; this suggests that $Ni(CO)_4^{2+}$ is less easily formed than the other observed ions. For $^{58}Ni(^{12}CO)_2^{2+}$, a peak at $114/+2 = 57$ a.m.u. does in fact appear. Likewise, the presence of $^{58}Ni(^{12}CO)^{2+}$ is indicated by a peak at $86/+2 = 43$ a.m.u. Like the $+1$ ions, the $+2$ ions also give characteristic isotopic patterns.

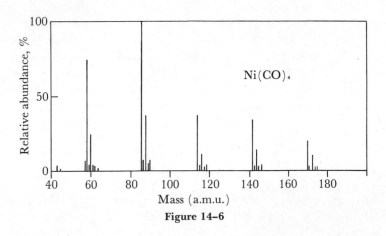

Figure 14–6

By such an analysis as we have made here for Ni(CO)$_4$, it should be possible to account for all the peaks in the mass spectrum of [1,3,5–C$_6$H$_3$(CH$_3$)$_3$]Mo(CO)$_3$. The characteristic distribution of Mo isotopes allows one to easily identify fragments containing Mo.

Isotope	Natural Abundance
^{92}Mo	15.86 per cent
^{94}Mo	9.12 per cent
^{95}Mo	15.70 per cent
^{96}Mo	16.50 per cent
^{97}Mo	9.45 per cent
^{98}Mo	23.75 per cent
^{100}Mo	9.62 per cent

Since there are many more C atoms in this compound, the abundance of ions containing one ^{13}C will also be higher than observed for Ni(CO)$_4$. It should be possible to assign almost all of the peaks in the spectrum to ions derived from the parent molecule. The percentage natural abundances for the isotopes of Cr and other elements are listed in Appendix 7. Use these values in interpreting the mass spectrum of (C$_6$H$_6$)Cr(CO)$_3$.

The amount of sample required for the determination of a mass spectrum may be invisibly small. Its size is usually limited by the operator's ability to handle the material. The mass spectrometer is a very expensive instrument, and a skilled technician is required to operate it. Because one may not be available for use with your sample, a mass spectrum of [1,3,5–C$_6$H$_3$(CH$_3$)$_3$]Mo(CO)$_3$, which we have recorded (Figure B), is given in the section on NMR and Mass Spectra at the end of the book. Fully account for the peaks in the spectrum.

REPORT

A. [1,3,5–C$_6$H$_3$(CH$_3$)$_3$]Mo(CO)$_3$

Include the following:

1. Percentage yield of [1,3,5–C$_6$H$_3$(CH$_3$)$_3$]Mo(CO)$_3$ and discussion of factors responsible for the low yield.
2. Infrared spectrum and the assignment of absorptions to C—O, C—H, and aromatic vibrations.
3. NMR spectrum and interpretation. Compare it with that of mesitylene in CHCl$_3$, which has absorptions at -2.25 and -6.78 ppm with relative intensities of 9 to 3, respectively.
4. Mass spectrum and the assignment of all peaks ranging from the parent ion, [1,3,5–C$_6$H$_3$(CH$_3$)$_3$]Mo(CO)$_3$$^+$, down to fragments having an m/e value of 90.

B. $(C_6H_6)Cr(CO)_3$

Include the following:

1. Percentage yield of $(C_6H_6)Cr(CO)_3$ and discussion of factors responsible for the low yield. Compare these factors with those in A.1.
2. Melting point.
3. Infrared spectrum and assignment of absorptions to C—O, C—H, and aromatic vibrations of $(C_6H_6)Cr(CO)_3$ and $Cr(CO)_6$.
4. NMR spectrum and comparison with free benzene.
5. Mass spectrum and ion assignments to all peaks.

QUESTIONS

1. If the reaction of $Mo(CO)_6$ with mesitylene were conducted in the presence of air, what would probably be the decomposition products?
2. How would you expect the C—O stretching frequencies in the compounds $(C_6H_6)Mo(CO)_3$, $[1,3,5-C_6H_3(CH_3)_3]Mo(CO)_3$, and $[C_6(CH_3)_6]Mo(CO)_3$ to vary and why?
3. $[1,3,5-C_6H_3(CH_3)_3]Mo(CO)_3$ reacts with $P(OCH_3)_3$ to form *cis*-Mo$(CO)_3[P(OCH_3)_3]_3$ according to the rate law:

$$Rate = k[(1,3,5-C_6H_3(CH_3)_3)Mo(CO)_3][P(OCH_3)_3]$$

   Postulate a mechanism for the reaction. Would you expect the same reaction with PF_3 to proceed slower or faster than that observed for $P(OCH_3)_3$ and why?
4. Outline a method for establishing the composition of $[1,3,5-C_6H_3(CH_3)_3]Mo(CO)_3$ by analyzing it for its percentage molybdenum content.
5. How would you determine in an afternoon's experiment whether or not your isolated $[1,3,5-C_6H_3(CH_3)_3]Mo(CO)_3$ (or $(C_6H_6)Cr(CO)_3$) was pure?
6. Considering only the equilibrium associated with reaction (1), which method of preparation of $[1,3,5-C_6H_3(CH_3)_3]Mo(CO)_3$ would give the higher yield: (1) as performed in the experiment or (2) in a bomb? Explain.
7. If you wished to carry out the following reaction, $(C_6H_6)Cr(CO)_3 + 1,3,5-C_6H_3(CH_3)_3 \rightarrow [1,3,5-C_6H_3(CH_3)_3]Cr(CO)_3 + C_6H_6$, what reaction conditions (solvent, temperature, amounts of reactants, etc.) would you use? Would you expect the equilibrium constant for the reaction to be greater or less than 1?
8. If you wished to carry out the reverse of reaction (1), $[1,3,5-C_6H_3(CH_3)_3]Mo(CO)_3 + 3CO \rightarrow Mo(CO)_6 + 1,3,5-C_6H_3(CH_3)_3$, what reaction conditions and techniques would you use?
9. A trap may be added to the apparatus in Figure 14–1 to prevent Hg from being drawn back into the reaction solution on cooling the reaction flask. Make a drawing that includes this modification.

INDEPENDENT STUDIES

A. Prepare the (Arene)Mo(CO)$_3$ complexes of C$_6$H$_5$N(CH$_3$)$_2$ and/or C$_6$(CH$_3$)$_6$. Compare their ir and nmr spectra with those of the mesitylene complex.

B. Prepare and characterize (π-cycloheptatriene)Mo(CO)$_3$. (F. A. Cotton, J. A. McCleverty, and J. E. White, *Inorganic Syntheses, Vol. 9*, McGraw-Hill, New York, 1967, p. 121.)

C. Prepare and characterize *cis*-[(CH$_3$O)$_3$P]$_3$Mo(CO)$_3$ obtained from the reaction of [1,3,5–C$_6$H$_3$(CH$_3$)$_3$]Mo(CO)$_3$ with P(OCH$_3$)$_3$. (A. Pidcock, J. D. Smith, and B. W. Taylor, *J. Chem. Soc. (A)*, 872 (1967). F. Zingales, A. Chiesa, and F. Basolo, *J. Amer. Chem. Soc.*, *88*, 2707 (1966).)

D. Analyze (C$_6$H$_6$)Cr(CO)$_3$ or [1,3,5–C$_6$H$_3$(CH$_3$)$_3$]Mo(CO)$_3$ for its percentage Cr or Mo content.

E. Prepare and characterize (C$_6$H$_6$)$_2$Cr. (E. O. Fischer, *Inorganic Syntheses, Vol. 6*, 132 (1960). P. L. Timms, *J. Chem. Educ.*, *49*, 782 (1972).)

F. Prepare and characterize a salt of (mesitylene)$_2$Fe^{2+}. (J. F. Helling and D. M. Braitsch, *J. Amer. Chem. Soc.*, *92*, 7207 (1970). J. F. Helling, S. L. Rice, D. M. Braitsch, and T. Mayer, *J. Chem. Soc., Chem. Commun.*, 930 (1971).)

REFERENCES

(Arene)M(CO)$_3$

D. M. Adams, R. E. Christopher, and D. C. Stevens, *Inorg. Chem.*, *14*, 1562 (1975). Infrared and Raman studies of (C$_6$H$_6$)Cr(CO)$_3$.

R. J. Angelici, *J. Chem. Educ.*, *45*, 119 (1968).

M. F. Bailey and L. F. Dahl, *Inorg. Chem.*, *4*, 1298, 1314 (1965). X-ray structural studies of (C$_6$H$_6$)Cr(CO)$_3$ and [C$_6$(CH$_3$)$_6$]Cr(CO)$_3$.

R. V. Emanuel and E. W. Randall, *J. Chem. Soc. (A)*, 3002 (1969). Proton nmr spectra of (Arene)Cr(CO)$_3$ compounds.

E. O. Fischer, K. Ofele, H. Essler, W. Fröhlich, J. P. Mortensen, and W. Semmlinger, *Chem. Ber.*, *91*, 2763 (1958). Preparation of (Arene)M(CO)$_3$ compounds.

R. D. Fischer, *Chem. Ber.*, *93*, 165 (1960). Infrared spectra of (Arene)Cr(CO)$_3$ compounds.

G. Klopman and K. Noack, *Inorg. Chem.*, *7*, 579 (1968). NMR spectra of (Arene)Cr(CO)$_3$.

B. Nicholls and M. C. Whiting, *J. Chem. Soc.*, 551 (1959). Preparation of (Arene)M(CO)$_3$ compounds.

S. Pignataro and F. P. Lossing, *J. Organometal. Chem.*, *10*, 531 (1967). Mass spectrum of (C$_6$H$_6$)Cr(CO)$_3$.

Techniques

R. L. Augustine, *Catalytic Hydrogenation*, Marcel Dekker, New York, 1965, p. 3. Description of apparatus and techniques used for high pressure bomb reactions.

M. I. Bruce in *Advances in Organometallic Chemistry, Vol. 6*, F. G. A. Stone and R. West, Eds., Academic Press, New York, 1968, p. 273. Mass spectra of organometallic compounds.

R. S. Drago, *Physical Methods in Inorganic Chemistry*, Reinhold Publishing Corp., New York, 1965, Chapters 7, 8, and 12. Basic treatment of infrared and nmr spectroscopy and mass spectrometry; applications to inorganic compounds.

S. Herzog, J. Dehnert, and K. Lühder in *Technique of Inorganic Chemistry, Vol. VII*, H. B. Jonassen and A. Weissberger, Eds., Interscience Publishers, New York, 1968, p. 119.

Description of glassware which may be used to carry out syntheses in an inert atmosphere without using a dry box.

L. M. Jackman, *Applications of Nuclear Magnetic Resonance Spectroscopy in Organic Chemistry,* Pergamon Press, New York, 1959. An elementary introduction to the theory and interpretation of nmr.

R. B. King, *Organometallic Syntheses, Vol. I.,* Academic Press, New York, 1965. Techniques of synthesis and characterization of organometallic compounds. A number of preparations of various organometallic compounds of the transition metals.

R. W. Kiser, *Introduction to Mass Spectrometry and Its Applications,* Prentice-Hall, Englewood Cliffs, New Jersey, 1965. Theoretical and practical aspects of the technique.

V. I. Komarewsky and C. H. Riesz in *Technique of Organic Chemistry, Vol. II,* A Weissberger, Ed., Interscience Publishers, New York, 1948. Describes different types of high pressure autoclaves (bombs).

M. L. Maddox, S. L. Stafford, and H. D. Kaesz in *Advances in Organometallic Chemistry, Vol. 3,* F. G. A. Stone and R. West, Eds., Academic Press, New York, 1965, p. 1. Tabulation and interpretation of nmr spectra of organometallic compounds.

J. L. Margrave and R. B. Polansky, *J. Chem. Educ., 39,* 335 (1962). H. M. Bell, *J. Chem. Educ., 51,* 548 (1974). Calculations of relative abundances of molecules with different isotopes for mass spectrometry.

J. D. Roberts, *Nuclear Magnetic Resonance: Applications to Organic Chemistry,* McGraw-Hill, New York, 1959. An elementary introduction to theory and interpretation of nmr.

D. F. Shriver, *The Manipulation of Air-Sensitive Compounds,* McGraw-Hill, New York, 1969. Excellent coverage of techniques for handling air-sensitive compounds.

R. M. Silverstein and G. C. Bassler, *Spectrometric Identification of Organic Compounds,* 2nd Ed., John Wiley and Sons, New York, 1967, Chapters 2, 3, and 4. A practical introduction to infrared, nmr, and mass spectrometry.

M. Tsutsui, Ed., *Characterization of Organometallic Compounds, Vols. 1 and 2,* Interscience Publishers, New York, 1969. A very useful general reference to instrumental methods of characterizing organometallic compounds.

Organometallic Chemistry

M. L. H. Green, *Organometallic Compounds, Vol. 2,* Methuen and Co. Ltd., London, 1968. An extensive coverage of the organometallic chemistry of the transition metals.

P. L. Pauson, *Organometallic Chemistry,* St. Martin's Press, New York, 1967. An introductory survey of the subject.

B. L. Shaw and N. I. Tucker, *Organo-Transition Metal Compounds and Related Aspects of Homogeneous Catalysis,* Pergamon Press, New York, 1973. A compact survey of organometallic chemistry.

W. E. Silverthorn, *Advances in Organometallic Chemistry, Vol. 13,* F. G. A. Stone and R. West, Eds., Academic Press, New York, 1975, p. 47. A review of arene complexes of the transition metals.

$\left[C_5H_5Fe(CO)_2\right]_2$ and $C_5H_5Fe(CO)_2CH_3$

Note: This experiment requires a reaction period of 8 to 10 hours.

Modern organometallic chemistry of transition metals began with the synthesis of the cyclopentadienyl complex ferrocene, $(C_5H_5)_2Fe$, in 1951. Since that time cyclopentadienyl complexes of a wide variety of metals, including all of the transition elements, have been prepared. Those to be synthesized in this experiment also contain a cyclopentadienyl group π-bonded to Fe:

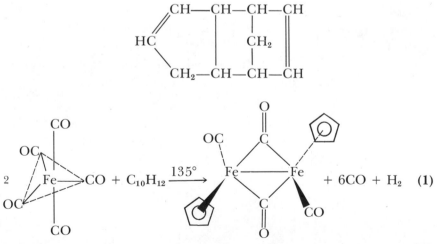

The first compound is obtained from the reaction of $Fe(CO)_5$ with dicyclopentadiene, the Diels-Alder dimer adduct of cyclopentadiene

The product, $\left[C_5H_5Fe(CO)_2\right]_2$, has a structure that contains some unusual features. First, it contains both terminal and bridging CO groups. As expected, the bridging CO groups resemble organic carbonyl,

$\diagdown$
$\diagup C = O$, groups more than do the terminal CO's. For example, the C–O

stretching frequencies of bridging CO groups ($\sim$1750–1875 cm^{-1}) occur at lower frequencies than those of the terminal CO's ($\sim$1875–2150 cm^{-1}). Second, the diamagnetism of the compound and the short Fe–Fe distance indicate the presence of a metal-metal bond.

The two Fe atoms and the bridging CO groups all lie in the same plane, while the C_5H_5 rings and terminal CO groups lie above and below that plane. Only the isomer with the C_5H_5 rings (and CO groups) *trans* to each other across the plane is shown in equation (1). In solution, both the *trans* and *cis* isomers are known to be in rapid equilibrium with each other and with a small amount of the non-bridged structure:

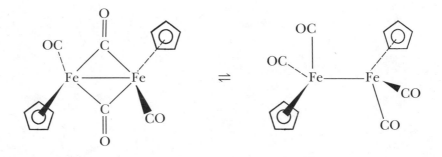

Like many organometallic compounds that contain metal-metal bonds, they may be reduced with metallic sodium to the monomeric anions. Thus, sodium amalgam reduces tetrahydrofuran (THF) solutions of $\left[C_5H_5Fe(CO)_2\right]_2$ to the very air sensitive $C_5H_5Fe(CO)_2{}^-$.

$$\left[C_5H_5Fe(CO)_2\right]_2 + 2Na/Hg \xrightarrow{\text{THF}} 2\ C_5H_5Fe(CO)_2{}^- + 2\ Na^+$$

This anion is a strong nucleophile and will, like numerous organic nucleophiles, displace halides from alkyl halides. The particular displacement reaction that will be conducted in this experiment is that on CH_3I:

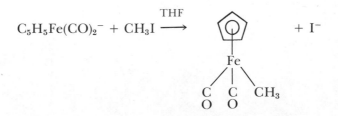

The product, C$_5$H$_5$Fe(CO)$_2$CH$_3$, is a volatile orange solid (m.p., 78° to 82°C), which decomposes in air at a significant rate.

The nucleophilicity of C$_5$H$_5$Fe(CO)$_2$⁻ facilitates the displacement of halides in other types of organic compounds also. For example,

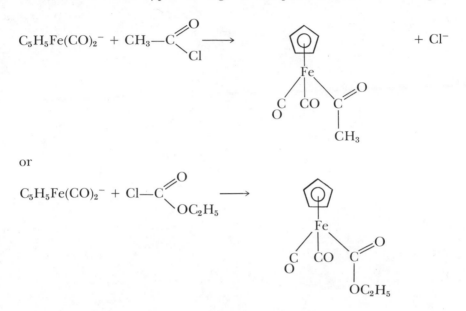

or

This last compound is rather unusual in that the ester carbonyl group may be converted to a terminal metal-carbonyl ligand on reaction with HCl:

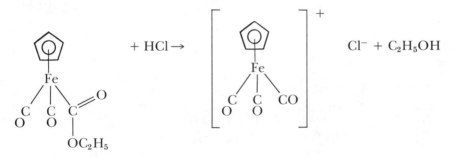

These latter two reactions have been used to prepare one of the few known compounds that contain the thiocarbonyl, CS, ligand, analogous to CO.

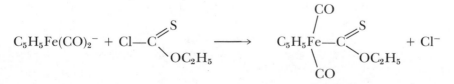

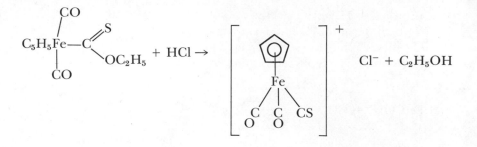

The cationic $C_5H_5Fe(CO)_3^+$ is particularly susceptible to attack by nucleophiles. The nature of the product may be quite varied, however. For example, where the nucleophile, R^-, is $C_2H_5O^-$, RNH^-, or $C_6F_5^-$, attack occurs at the C atom of a CO group to give

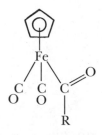

On the other hand, such nucleophiles, X^-, as Cl^-, Br^-, I^-, $C{\equiv}N^-$, $N{=}C{=}O^-$, and $N{=}C{=}S^-$ attack with displacement of a CO group. The product in these cases is

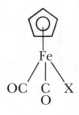

The reasons for the formation of different types of products with different nucleophiles are not clear.

While this experiment is concerned, at least in part, with the reduction of $\left[C_5H_5Fe(CO)_2\right]_2$ to $C_5H_5Fe(CO)_2^-$, numerous other organometallic compounds containing metal-metal bonds may likewise be converted to anions upon reaction with sodium amalgam. Some examples are

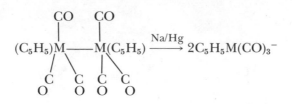

where M = Cr, Mo, or W

$$(OC)_5Mn\!\!-\!\!Mn(CO)_5 \xrightarrow{\text{Na/Hg}} 2Mn(CO)_5{}^-$$

$$(OC)_4Co\!\!-\!\!Co(CO)_4 \xrightarrow{\text{Na/Hg}} 2Co(CO)_4{}^-$$

These anions are also nucleophiles and will undergo many of the reactions noted above for $C_5H_5Fe(CO)_2{}^-$.

EXPERIMENTAL PROCEDURE

Iron pentacarbonyl is a toxic yellow liquid with a boiling point (103°) that is about the same as that of water. It has a readily recognizable musty odor and should always be handled in an efficient *hood*. The relatively high volatility of $Fe(CO)_5$ makes it particularly dangerous, and it should be stored in the hood. The evolution of CO in the preparation of $\left[C_5H_5Fe(CO)_2\right]_2$ is further reason for conducting the reaction in the *hood*.

[C₅H₅Fe(CO)₂]₂

This reaction *must be carried out in a hood*. Assemble the reaction vessel as shown in Figure 15–1. Flush the system for 5 minutes with a rapid stream of nitrogen. With the nitrogen stream still flowing, remove the thermometer, and add 14.6 g (70.5 mmoles, 10 ml) of $Fe(CO)_5$ and 60 g (455 mmoles, 64 ml) of good grade dicyclopentadiene ($C_{10}H_{12}$) to the round-bottom flask. To minimize exposure to $Fe(CO)_5$, one should use a syringe to measure out and introduce the $Fe(CO)_5$ into the flask. The constant stream of N_2 will prevent air from entering the flask while the reactants are being added. After replacing the thermometer, turn off the nitrogen flow both at the tank and with the inlet stopcock on the flask.

Heat the reaction mixture at 135°C for 8 to 10 hours. It is important that the temperature not be allowed to go below 130° or above 140°. Below 130° very little reaction occurs, and above 140° substantial decomposition to pyrophoric, finely divided metallic iron is produced.

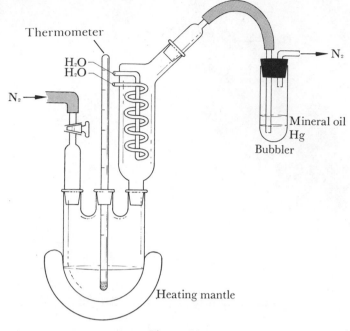

Figure 15-1

After the reaction period, turn on a slow stream of nitrogen and allow the mixture to slowly cool to room temperature. The nitrogen flow will prevent the Hg and mineral oil from being drawn into the reaction mixture as it cools. The red-violet crystals of $[C_5H_5Fe(CO)_2]_2$ are suction-filtered (Figure 1-1), washed at least four times with 20 ml portions of pentane, and sucked dry on the frit. The compound may be purified by dissolving it in a minimum of $CHCl_3$, filtering it, and adding an equal volume of hexane. Slowly evaporate the solvent under a water-aspirator vacuum (Figure 6-3) to about half the original volume. The more volatile $CHCl_3$ will evaporate readily, causing the product to precipitate from the hexane. Generally this purification is not necessary for the synthesis of $C_5H_5Fe(CO)_2CH_3$. Calculate the percentage yield. Determine its melting point and measure its infrared spectrum in $CHCl_3$ solvent (see Experiment 14, p. 136). Its proton nmr spectrum (Experiment 14, p. 137) exhibits a single sharp resonance at -4.64 ppm downfield from TMS.

$C_5H_5Fe(CO)_2CH_3$

Dry tetrahydrofuran (THF), required as a solvent in this preparation, may be obtained from the more or less wet commercially available

THF as described in Experiment 6, p. 58. A good commercial grade of THF may be used directly without drying. The reaction will be carried out in a 250 ml 3-neck round-bottom flask with a 2 mm or larger bore stopcock fused to the bottom, as shown in Figure 15–2.

Assemble the apparatus (lubricate the bearing with a few drops of glycerin), and flush with a stream of nitrogen for 5 minutes as was done in the previous preparation. With the nitrogen flowing, remove the condenser and pour 15 ml of mercury into the flask. Weigh 1.0 g (43 mmoles) of freshly cut sodium metal, and with stirring add it to the mercury in ~0.2 g portions. To avoid reaction with air, it is convenient to put the 1 g piece in a crystallizing dish and cover it with a hydrocarbon solvent that has a boiling point between 80° and 100°C. It may then be cut into the 0.2 g pieces without exposure to air. Dispose of waste Na as indicated in Experiment 4, p. 42. Since the reaction of Na and Hg is highly exothermic, the addition of Na to the Hg will be accompanied by a relatively loud hissing sound.

After the amalgamation is complete and the flask has cooled to room temperature, add 50 ml of dry THF and stir for a few minutes.

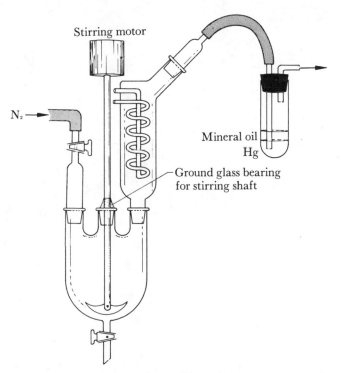

Figure 15–2

Then add 2.5 g (7 mmoles) of $[C_5H_5Fe(CO)_2]_2$. Replace the condenser and flush the system for a few minutes with nitrogen. Turn off the nitrogen. Adjust the speed of the stirrer with a rheostat until the amalgam is being *vigorously* mixed with the THF solution. After 15 minutes, note the color change and stop the stirring. Drain off the mercury amalgam through the stopcock at the bottom of the flask. (At the end of the experiment, the mercury may be recovered from the amalgam by first washing it with ethanol to destroy any sodium as $NaOC_2H_5$ and H_2. Then wash with water; if any Na remains in the Hg, its reaction with water is vigorous but not explosive [see Experiment 4, p. 42]. After separation from the water, the mercury is sufficiently dry for reuse in this type of reaction.)

With the N_2 flowing, remove the condenser and pour 2 ml (32 mmoles) of CH_3I into the solution of $Na^+[C_5H_5Fe(CO)_2{}^-]$ and note the color change. The reaction is exothermic. After stirring 15 minutes at room temperature, replace the stirrer and condenser with ground glass stoppers. Evaporate the solution to dryness under a water-aspirator vacuum at room temperature (Figure 6–3). Detach the vacuum tubing

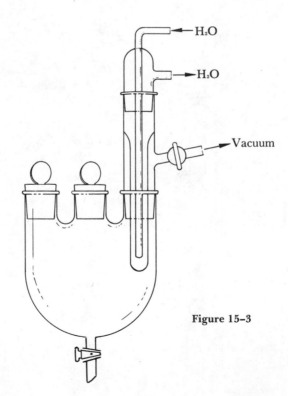

Figure 15–3

and admit nitrogen into the flask. Remove the nitrogen inlet and insert a sublimation probe as in Figure 15–3. Immediately evacuate the apparatus on the utility high vacuum line, but do not leave it open to the dynamic vacuum for more than a few minutes because the volatile C$_5$H$_5$Fe(CO)$_2$CH$_3$ will be drawn off into the trap of the vacuum line. Close the stopcock on the sublimation probe and cool the cold finger by circulating cold water through it. The product at room temperature will sublime slowly onto the cold finger. Orange waxy C$_5$H$_5$Fe(CO)$_2$CH$_3$ may be handled for short periods in air without serious decomposition. It should be stored under vacuum or nitrogen in a container like that shown in Figure 6–4. Scrape the product off the probe and determine the percentage yield. (With somewhat poorer yields, the product may be sublimed by transferring the reaction mixture to a sublimation apparatus such as that in Figure 11–3.)

Determine its melting point in a capillary tube sealed at both ends to guard against air oxidation during the determination. Record an infrared spectrum of the compound in CS$_2$ solution (see Experiment 14, p. 136). Measure the proton nmr spectrum of a CS$_2$ solution of the compound. Since C$_5$H$_5$Fe(CO)$_2$CH$_3$ solutions decompose quite rapidly in air, both the infrared and nmr spectra should be recorded immediately after the solutions are prepared. If an nmr spectrometer is not available, interpret the spectrum (Figure C) of C$_5$H$_5$Fe(CO)$_2$CH$_3$, which is given in the section on NMR and Mass Spectra near the end of this book.

REPORT

Include the following:
1. Percentage yield and melting point of $\left[\text{C}_5\text{H}_5\text{Fe(CO)}_2\right]_2$.
2. Infrared spectrum and its interpretation in terms of the structure of $\left[\text{C}_5\text{H}_5\text{Fe(CO)}_2\right]_2$.
3. Yield and melting point of C$_5$H$_5$Fe(CO)$_2$CH$_3$.
4. Infrared and proton nmr spectra and their interpretation in terms of the structure of C$_5$H$_5$Fe(CO)$_2$CH$_3$.

QUESTIONS

1. Elemental analysis is still a standard method of characterizing new compounds. Outline a procedure for analyzing your sample of $\left[\text{C}_5\text{H}_5\text{Fe(CO)}_2\right]_2$ for its percentage iron content.
2. The preparations in this experiment were conducted in a nitrogen atmosphere. What products would probably have formed if the reactions had been simply carried out in air?
3. In terms of the bonding in $\left[\text{C}_5\text{H}_5\text{Fe(CO)}_2\right]_2$, account for the diamagnetism of this complex.

4. What reactions would you use to prepare C_5H_5Fe—C (with CO, CO ligands and a $C=O$, C_6H_5 acyl group)?

5. Draw the structure of the *cis* isomer of $[C_5H_5Fe(CO)_2]_2$.

6. What is the purpose of using mineral oil on mercury in the bubblers in Figures 15–1 and 15–2 rather than using only mineral oil or mercury?

7. While both *cis* and *trans* isomers of $[C_5H_5Fe(CO)_2]_2$ are known to be present in solution, proton nmr spectra of its solutions at 25°C show only a sharp singlet. Account for the fact that only a singlet is observed.

8. Would the Fe–CH$_3$ bond in $C_5H_5Fe(CO)_2CH_3$ be more or less ionic than the Fe–Cl bond in the related $C_5H_5Fe(CO)_2Cl$? What experimental evidence would you need to support your conclusion?

INDEPENDENT STUDIES

A. From the reaction of $C_5H_5Fe(CO)_2CH_3$ with $P(C_6H_5)_3$, isolate and characterize the acyl product, $C_5H_5Fe(CO)[P(C_6H_5)_3]C(=O)CH_3$. (J. P. Bibler and A. Wojcicki, *Inorg. Chem.*, 5, 889 (1966). M. Green and D. J. Westlake, *J. Chem. Soc. (A)*, 367 (1971).)

B. Determine the mass spectra of $[C_5H_5Fe(CO)_2]_2$ and/or $C_5H_5Fe(CO)_2$-CH$_3$ and make assignments to all peaks.

C. Analyze $[C_5H_5Fe(CO)_2]_2$ for its percentage iron content.

D. Prepare and characterize $C_5H_5Fe(CO)_2Cl$ or $C_5H_5Fe(CO)_2I$ obtained from $[C_5H_5Fe(CO)_2]_2$. (P. M. Treichel and J. P. Stenson, *Inorganic Syntheses, 12*, 35 (footnote) (1970). R. B. King and F. G. A. Stone, *Inorganic Syntheses, 7*, 110 (1963).)

E. Prepare and characterize the metal-metal bonded $[C_5H_5Fe(CO)_2]_2$-SnCl$_2$ obtained from $[C_5H_5Fe(CO)_2]_2$ and SnCl$_2$·2H$_2$O. (F. Bonati and G. Wilkinson, *J. Chem. Soc.*, 179 (1964).)

REFERENCES

[C$_5$H$_5$Fe(CO)$_2$]$_2$ and C$_5$H$_5$Fe(CO)$_2$CH$_3$

R. F. Bryan and P. T. Greene, *J. Chem. Soc. (A)*, 3064 (1970). X-ray structural study of *trans*-[C$_5$H$_5$Fe(CO)$_2$]$_2$.

R. F. Bryan, P. T. Greene, M. J. Newlands, and D. S. Field, *J. Chem. Soc. (A)*, 3068 (1970). X-ray structural study of *cis*-[C$_5$H$_5$Fe(CO)$_2$]$_2$.

R. B. King, *Organometallic Syntheses, Vol. 1*, Academic Press, New York, 1965, pp. 114 and 151.

T. S. Piper and G. Wilkinson, *J. Inorg. Nucl. Chem., 3*, 104 (1956).

Techniques and Organometallic Chemistry

See references at the end of Experiment 14.

Chromatography of Ferrocene Derivatives

Since the initial preparations of ferrocene, $Fe(C_5H_5)_2$, in 1951, numerous investigators have examined the reactions of this compound to determine whether the cyclopentadienyl rings are similar to benzene in their chemical reactivity. In fact, many substitution reactions on the cyclopentadienyl rings do occur, and ferrocene usually undergoes these reactions more readily than does benzene. These observations have been interpreted to indicate that the cyclopentadienyl rings in ferrocene are "more aromatic" than benzene. Regardless of how one defines aromaticity, it is at least clear that ferrocene readily undergoes electrophilic substitution. One such substitution reaction is that of acetylation in the presence of a Friedel-Crafts catalyst.

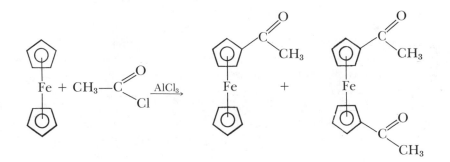

The role of the Lewis acid $AlCl_3$ is presumably to participate in the generation of the electrophile, $CH_3C\equiv O^+$.

$$CH_3C{\overset{O}{\underset{Cl}{\diagup\!\!\backslash}}} + AlCl_3 \rightleftharpoons CH_3C\equiv O^+ + AlCl_4^-$$

Whether the mono- or diacetyl product is obtained is determined by the amounts of reactants and the conditions of the reaction.

In this experiment, the acetylation of ferrocene will be carried out under milder conditions to yield primarily the monoacetyl product.

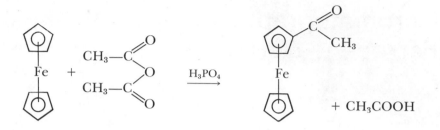

The electrophile, $CH_3C\equiv O^+$, is probably produced by the reaction,

$$(CH_3CO)_2O + H_3PO_4 \rightleftharpoons CH_3C\equiv O^+ + CH_3COOH + H_2PO_4^-$$

Since H_3PO_4 produces a relatively small concentration of $CH_3C\equiv O^+$ compared to that generated in the reaction of CH_3COCl with $AlCl_3$, only small amounts of the monoacetyl derivative are converted to the disubstituted compound.

Thin Layer Chromatography

While the chemistry of ferrocene is novel and extensive, it is left to the reader to pursue this topic in further depth through the references given at the end of this experiment. One of the primary purposes of this experiment is to illustrate standard chromatographic techniques that are used in the separation of pure compounds from complex reaction mixtures. In the acetylation reaction of ferrocene, a solid mixture of primarily $Fe(C_5H_5)_2$ and $Fe(C_5H_5)(C_5H_4COCH_3)$ is obtained. The remainder of this experiment is concerned with chromatographic methods of separating these two complexes. Thin layer chromatography (TLC) will be used to explore what chromatographic conditions are necessary to separate the compounds; employing these conditions, column chromatography will be utilized to separate the pure ferrocenes on a larger scale.

Separations effected by both TLC and column chromatography are based on the tendency of molecules to adsorb to certain adsorbents. The adsorbents most frequently used in TLC and column chromatography

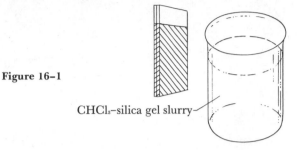

Figure 16–1

CHCl₃–silica gel slurry

are silica gel and alumina. (See Appendix 8.) In TLC, a thin layer (0.1 to 2 mm thick) of the adsorbent is fixed to the surface of a flat plate. While there are many variations of this technique, microscope slides (2.5 × 7.5 cm) are commonly employed and will be used in this experiment as the plate. Two microscope slides are placed back to back and dipped into a stirred chloroform or chloroform-methanol slurry (Figure 16–1) of the adsorbent (silica gel in this experiment).

The slides are separated and allowed to dry in air for several minutes. A small amount of the sample to be separated is dissolved in a small volume of a suitable solvent. With a capillary, a spot (3 to 5 mm in diameter) of the solution is placed on the silica gel 8 to 10 mm from the end of the slide. Allow it to dry and then add more of the solution to the same spot. The spot should be kept small for maximum separation of the components. This slide is then placed in a weighing bottle that contains a little solvent (~4 mm deep) (Figure 16–2). The solvent level should be lower than the spot on the slide. Close the bottle and allow the solvent to rise up the silica gel, and if the proper solvent was chosen the mixture will begin to separate and move up the plate behind the solvent front. If different compounds move up the silica gel at different rates they will be separated.

When the solvent has moved about three-fourths of the way up the silica gel, the plate should be removed. If the compounds in the mixture are colored, it will be obvious if a separation occurs. If one or

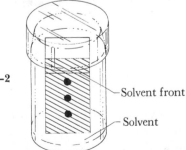

Figure 16–2

Solvent front

Solvent

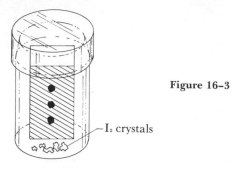

Figure 16–3

I₂ crystals

more of the compounds are colorless, their locations on the slide can usually be established by placing the air-dried slide in a weighing bottle containing a few crystals of iodine (Figure 16–3). The iodine sublimes and adsorbs in the areas where the compounds are located. Thus, dark brown spots on the plate indicate the locations of the components of the original sample. If the solvent that was chosen for the separation does not separate the components of the mixture, another more or less polar solvent should be tried until a solvent that gives a separation is found. The solvent that does give a separation will then be used in the larger scale isolation of the compounds by column chromatography.

The separation of mixtures into their components by TLC is governed by the nature of the adsorbent and the solvent. The forces that attract a compound to the silica gel on a TLC plate are assumed to be largely polar. The separation of a mixture of compounds therefore depends to a large extent on the differences in adsorption tendencies of the components. The nature of this adsorption is very complex and poorly understood. Silica gel, for example, is largely SiO_2 or hydrated forms of SiO_2, sometimes written as $Si(OH)_4$ or $SiO_2 \cdot xH_2O$. It does, however, contain significant amounts of other inorganic salts whose amounts vary from one silica gel preparation to another.

When silica gel is heated in strongly acid or basic solutions, it acquires acidic or basic properties and even can act to some extent like a cation or anion exchange column. Acid-treated silica gel strongly adsorbs (or binds) basic compounds such as amines, whereas base-treated silica gel adsorbs acidic compounds. For silica gels that are neutral (i.e., when it is washed with water, the wash is approximately neutral) there are even differences in adsorption depending upon the water content of the silica gel. The most active form is obtained by heating under vacuum at 160°C for 4 hours. Most of the water is driven off the silica gel leaving sites where other polar molecules might adsorb. Less strongly adsorbing silica gel can be prepared by adding 10, 15, or 20 per cent H_2O to occupy some of the adsorption sites. By altering the water content, it is therefore possible to alter the degree of adsorption of a compound on a TLC plate.

The extent of adsorption is also determined by the compound's solubility in the solvent and the affinity of the solvent for the adsorption sites. Polar solutes will adsorb strongly to the silica gel when a relatively nonpolar solvent is used. If a solvent of higher polarity is used, the solute will be more soluble in it, and the solvent will have a greater tendency to adsorb to the silica gel by displacing some of the solute molecules. Both of these factors favor less adsorption and faster migration up the TLC slide as the polarity of the solvent is increased.

The choice of solvent is not easy, but generally a solvent that dissolves the desired compound moderately well will allow the compound to move up the plate but not as fast as the solvent. It can only be hoped that the impurities do not move at the same rate as does the compound of interest. If the compound moves too rapidly, a less polar solvent should be tried, whereas if it moves too slowly a more polar solvent is appropriate. The trend of polarity of some common chromatographic solvents is shown as follows:

light petroleum, pentane, hexane, and the like
carbon tetrachloride
toluene
benzene
dichloromethane
chloroform
ethyl ether
ethyl acetate
acetone
ethanol
methanol
water

(increasing polarity ↓)

At this point it is probably obvious that the successful choice of adsorbent and solvent is an art that is learned largely by doing chromatographic separations. The references at the end of this experiment do offer, however, many hints on how to use these techniques more effectively.

Column Chromatography

Having established the solvent or solvent mixture that will separate the sample on TLC plates, it is hoped that the same solvent can be used to separate larger quantities of the sample on a silica gel chromatography column. Generally this is possible. It is necessary, however, to use a much larger silica gel particle size for column chromatography (80–200 mesh) than that used in TLC (finer than 200 mesh). This necessitates using a different source of silica gel, which may or may not have the same adsorption characteristics as that used in TLC. Fortunately, silica gels are frequently similar enough that a mixture can be separated on a column using the same solvent that was used successfully on the TLC plates. Initially, it is reasonable to assume that this is possible.

A simple burette may be used for the chromatography column (Figure 16–4). First a small glass wool plug is pushed to the bottom of the column, and a 5 mm layer of washed sand is added. Make up a slurry of the silica gel in the solvent to be used in the separation and pour it onto the sand. After draining the column until the solvent level is the same as the top of the silica gel, add a slurry of the sample and silica gel in a few milliliters of the solvent to the column. Then add a 5 mm layer of sand to the top of the silica gel. (Instead of adding the sample as a slurry, it may be dissolved in a small volume of the solvent and added to the column.) Then the liquid level is lowered to the top of the silica gel. Never allow the solvent level to fall below the top of the silica gel, for channels in the column will result, and solution will pass down the channels without properly percolating through the adsorbent. The eluting solvent is added to the column, and the elution of the compounds begins. The flow-rate will depend upon the separation, but slow flow-rates give better separations than do high flow-rates.

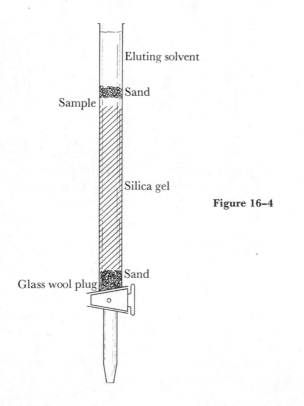

Figure 16–4

If the compounds are colored, it is easy to determine what fractions should be collected to obtain the products. If the compounds are colorless, fractions may have to be collected at regular volume intervals and examined for the presence of compounds spectrophotometrically or by other techniques. While the initial eluting solvent may elute one or more of the compounds, other compounds may require (as noted by TLC) more polar solvents to move them down the column. If a change in solvent is required, it is best to introduce it gradually by using first a mixture of the initial and the subsequent solvent and then finally the pure solvent. Sometimes a direct change of solvent involves a large evolution of heat when the new solvent adsorbs to the silica gel. This causes the solvent to expand, and channels are formed in the adsorbent that destroy the efficiency of the column.

The fractions eluted from the column that contain the desired compounds may simply be evaporated to dryness to give the pure compound. Evaporation sometimes does not give a crystalline solid, and recrystallization of the material usually gives a better looking product.

Column chromatography has numerous variations. Although silica gel and alumina ("Al_2O_3") are the most common adsorbents, many others have also been used. (See Appendix 8.) For materials that decompose at room temperature, chromatographic separations have been carried out in cooled, jacketed columns. Air-sensitive compounds have been chromatographed in an atmosphere of nitrogen or argon. All of these variations, however, are basically chromatography, and it is these basics that will be practiced in this experiment.

The separation of $Fe(C_5H_5)_2$ (m.p., 173–4°C) and $Fe(C_5H_5)(C_5H_4-COCH_3)$ (m.p., 83–5°C) is a relatively easy one. There is more than one solvent or solvent combination that will provide the desired separation. It is the student's task to determine at least one solvent system that will work. Having separated the compounds, the eluted products must then be identified by their melting points and their infrared (see Experiment 14, p. 136) and proton nuclear magnetic resonance (see Experiment 14, p. 137) spectra.

EXPERIMENTAL PROCEDURE

The reactant, $Fe(C_5H_5)_2$, and product, $Fe(C_5H_5)(C_5H_4COCH_3)$, are stable solids in both air and water at room temperature.

Acetylcyclopentadienyl-(cyclopentadienyl)-iron(II), $Fe(C_5H_5)(C_5H_4COCH_3)$

Add 1 ml of 85 per cent phosphoric acid dropwise with constant stirring to a mixture of 1.5 g (8.05 mmoles) of ferrocene and 5 ml (5.25

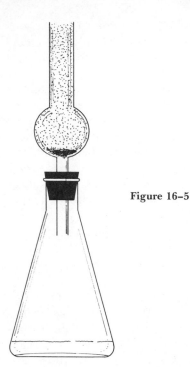

Figure 16–5

g, 87 mmoles) of acetic anhydride in a small Erlenmeyer flask. Protect the mixture with a $CaCl_2$ drying tube (Figure 16–5). Heat the reaction mixture on a steam bath (or in a boiling water bath) for 10 minutes. Then pour the mixture onto about 20 g of ice in a tall beaker. When the ice has melted, neutralize the mixture by adding solid $NaHCO_3$ until CO_2 is no longer evolved. (This will require a fairly large amount of $NaHCO_3$.) Cool the mixture in an ice bath for 30 minutes to ensure complete precipitation of the ferrocenes from solution. Suction-filter (Figure 1–1) the solid on a coarse frit and wash it with water until the filtrate is pale orange. Air-dry the solid on the frit for 15 minutes. This solid largely contains $Fe(C_5H_5)_2$ and $Fe(C_5H_5)(C_5H_4COCH_3)$, but other impurities are also present.

Thin Layer Chromatography, TLC

Using TLC, determine which solvent system will allow you to separate $Fe(C_5H_5)(C_5H_4COCH_3)$ from the product mixture by column chromatography.

Prepare the TLC plates (Figure 16–1) by placing two microscope slides (2.5 × 7.5 cm) back to back and dipping them into a stirred chloroform slurry of TLC silica gel (approximately 40 g of silica gel per 100 ml of $CHCl_3$). Withdraw them slowly and touch them to the edge of the container to allow them to drain. Separate the slides, remove excess silica gel on the edges with your fingers, and allow the slides to air-dry on a horizontal surface for about 5 minutes. Handle the slides carefully to keep the silica gel layer intact. (Keep the slurry bottle closed when not in use because silica is hygroscopic.)

With a very small portion of the ferrocene mixture, prepare a concentrated benzene solution for use in the TLC trial separations. A very small amount of this solution will be spotted on the silica gel 8 to 10 mm from the bottom of the microscope slide. Since the best separations are achieved when this spot is the smallest, the benzene solution should be applied with a very small capillary. It can be prepared by heating the middle of a melting point capillary tube over a low flame and quickly pulling the ends of the tube apart before the tube is sealed off. Scratch the capillary with a sharp file and break the tube into two applicators. Dip an applicator into the benzene solution of the mixture and touch it to the slide, giving a spot that is not larger than 4 mm in diameter. Allow the benzene to evaporate, and then make a second application of the solution to the same spot.

Fill five weighing bottles (Figure 16–2) with the trial solvents listed below to a depth of 4 mm or less, such that the level is below your spot when the microscope slide is inserted.
1. petroleum ether, 60°C to 70°C boiling fraction
2. benzene
3. ethyl ether
4. 10 per cent ethyl acetate and 90 per cent petroleum ether (by volume)
5. ethyl acetate

Insert a spotted slide into each bottle and replace the cover. Allow the solvent to rise about three-fourths of the way up the slide and then remove it and allow it to dry. Do this for each of the five slides, carefully recording what solvent was used for each slide. Although $Fe(C_5H_5)_2$ and $Fe(C_5H_5)(C_5H_4COCH_3)$ are colored, they may hardly be visible in the low concentrations on the plate. Their presence can be unequivo-cally established by placing each of the slides in a closed weighing bottle (Figure 16–3) containing a few crystals of iodine. The volatile iodine will adsorb to the plate where compounds are located.

Make a drawing in your research notebook of the location of the spots for each of the five attempts. To establish which spot is ferrocene, prepare another TLC slide, spotting it first with the benzene solution of the mixture and then in an adjacent position with a benzene solution of pure ferrocene. Develop the slide in one of the solvents that gave a good separation of spots, and establish which spot of the mixture is ferrocene by comparing it with the known ferrocene spot.

From the five TLC trials you should select a solvent for the column chromatographic separation. You may select a solvent in which one of the components moves rapidly and the other more slowly. Such a solvent should give a good separation on the column. Alternatively, one might choose an initial solvent in which only one component moves while the other remains at the starting point. The first component could then be washed off the column, and then a more polar solvent that moves both materials could be added to elute the other component. Regardless of your choice, it should be based on the separation achieved on the TLC slides.

Column Chromatography

A 25 ml burette (9 mm, i.d.) may be used (Figure 16-4). Push a wad of glass wool to the bottom of the burette with a glass rod. Then pour in enough washed sand to give a 5 mm layer. Make a slurry of silica gel (80–200 mesh) in the solvent that you initially plan to use in the separation, and pour it into the column until the silica gel column height is about 30 cm. (Keep the stock bottle of silica gel closed when not in use. Dry silica gel is very hygroscopic and will become deactivated if exposed to atmospheric humidity.) Drain the burette until the solvent level is the same as the silica gel level. (*Do not* allow the solvent level to drop below the top of the silica gel.) Then pour a slurry of silica gel and about 0.4 g of the ferrocene mixture in a few milliliters of the initial solvent onto the column. (Depending upon which solvent you choose, the sample may or may not dissolve. Even if it does not dissolve, a good separation should result if you have selected your solvents correctly.) Lower the solvent again at the top of the adsorbent and add another 5 mm layer of washed sand to prevent the silica gel bed from being disturbed when the eluting solvent is added. Then gently fill the burette with the initial eluting solvent, being careful not to agitate the silica gel. Carry out the elution, using a flow rate of approximately 1 drop per second. Since ferrocenes decompose somewhat in light, the column should be loosely wrapped with aluminum foil. The progress of the separation can be noted by momentarily opening the foil. The color of the two ferrocenes should allow you to collect separately solutions of $Fe(C_5H_5)_2$ and $Fe(C_5H_5)(C_5H_4COCH_3)$. Transfer each solution to a small filter-flask. Stopper the top of the flask and evaporate the solution to dryness under water aspiration (Figure 6-3). Weigh each of the solids. The tar that remains at the top of the column should be discarded along with the rest of the used silica gel.

The two colored products that are collected must be identified and characterized. Measure their melting points. Record the infrared spectrum of each compound in CCl_4 solution. Obtain the nmr spectrum of the two compounds in CCl_4 or benzene solution. [If an nmr spectrometer is not available, the spectrum (Figure D) of $Fe(C_5H_5)(C_5H_4COCH_3)$ at the back of this book should be interpreted. The nmr spectrum of $Fe(C_5H_5)_2$ taken in benzene solvent shows only one absorption at -4.02 ppm with respect to tetramethylsilane.] Having identified the $Fe(C_5H_5)$-

$(C_5H_4COCH_3)$, calculate the percentage yield of the product and the percentage recovery of unreacted ferrocene.

REPORT

Include the following:
1. Percentage yield of $Fe(C_5H_5)(C_5H_4COCH_3)$ and percentage recovery of $Fe(C_5H_5)_2$.
2. Melting points of $Fe(C_5H_5)(C_5H_4COCH_3)$ and $Fe(C_5H_5)_2$.
3. Infrared spectra of $Fe(C_5H_5)(C_5H_4COCH_3)$ and $Fe(C_5H_5)_2$. Point out differences and similarities.
4. NMR spectra of $Fe(C_5H_5)(C_5H_4COCH_3)$ and $Fe(C_5H_5)_2$ and interpretation.

QUESTIONS

1. Was $Fe(C_5H_5)(C_5H_4COCH_3)$ or $Fe(C_5H_5)_2$ eluted first? Why?
2. Why was the reaction to form $Fe(C_5H_5)(C_5H_4COCH_3)$ protected with a $CaCl_2$ drying tube?
3. What might be the tar that remained at the top of the chromatography column?
4. Account for the different chemical shifts of the H's in the acetylated ring of $Fe(C_5H_5)(C_5H_4COCH_3)$ as compared to that of the H's in the nonacetylated ring.
5. Write a complete mechanism for the formation of acetylferrocene by the reaction used in this experiment.
6. The rates at which ferrocene derivatives elute from a silica gel column depend upon any pre-treatment of the silica gel. Would acetylferrocene move down a column made of silica gel that had been heated at 150° under vacuum for 8 hours faster or slower than it would on a column using silica gel that had been sitting open in the laboratory for a few days? Explain.
7. What methods might be used to detect the elution of colorless compounds from a column?
8. A mixture of *cis* and *trans* isomers of the neutral complex $Cr(CO)_4$-$[P(C_6H_5)_3]_2$ is loaded onto a column and eluted with $CHCl_3$. Which isomer would elute first, and why?

INDEPENDENT STUDIES

A. Analyze the $Fe(C_5H_5)_2$–$Fe(C_5H_5)(C_5H_4COCH_3)$ reaction mixture for the amounts of each of these compounds, using *gas chromatography*. (O. E. Ayers, T. G. Smith, J. D. Burnett, and B. W. Ponder, *Anal. Chem.*, *38*, 1606 (1966).)
B. Measure the mass spectra of $Fe(C_5H_5)_2$ and/or $Fe(C_5H_5)(C_5H_4COCH_3)$ and make assignments to all peaks.

C. Separate the $Fe(C_5H_5)_2$–$Fe(C_5H_5)(C_5H_4COCH_3)$ reaction mixture, using *dry-column chromatography*. (J. C. Gilbert and S. A. Monti, *J. Chem. Educ., 50,* 369 (1973).)
D. Prepare and characterize ferricinium picrate, $\left[Fe(C_5H_5)_2{}^+\right]\left[C_6H_2O_7\text{-}N_3{}^-\right]$. (D. W. Johnson and G. W. Rayner-Canham, *J. Chem. Educ., 49,* 211 (1972).)
E. Synthesize and characterize nickelocene, $Ni(C_5H_5)_2$. (K. W. Barnett, *J. Chem. Educ., 51,* 422 (1974).)

REFERENCES

$Fe(C_3H_5)(C_5H_4COCH_3)$

R. E. Bozak, *J. Chem. Educ., 43,* 73 (1966).
R. J. Graham, R. V. Lindsey, G. W. Parshall, M. L. Peterson, and G. M. Whitman, *J. Am. Chem. Soc., 79,* 3416 (1957).

Chromatographic Techniques

J. M. Bobbitt, A. E. Schwarting, and R. J. Gritter, *Introduction to Chromatography,* Reinhold Publishing Corp., New York, 1968. An excellent, practical introduction to thin layer, column, and gas chromatography. Contains extensive general references and a list of chromatographic equipment suppliers (paperback).
L. F. Druding and G. B. Kauffman, *Coord. Chem. Rev., 3,* 409 (1968). Thin layer, column, and paper chromatography of coordination complexes.
G. Guiochon and C. Pommier, *Gas Chromatography in Inorganics and Organometallics,* Ann Arbor Science Publishers, Inc., Ann Arbor, Michigan, 1973. Excellent coverage of experimental techniques and the literature.
E. Heftmann, Ed., *Chromatography,* 2nd Ed., Reinhold Publishing Corp., New York, 1967. Column, paper, gas, ion exchange chromatography.
J. S. Keller, H. Veening, and B. R. Willeford, *Anal. Chem., 43,* 1516 (1971). Gas chromatographic separation of $(Arene)Cr(CO)_3$ complexes.
J. G. Kirchner, *Thin Layer Chromatography,* Interscience Publishers, New York, 1967.
E. Lederer and M. Lederer, *Chromatography, a Review of Principles and Applications,* 2nd Ed., Elsevier, New York, 1957.
D. C. Malins and J. C. Wekell, *J. Chem. Educ., 40,* 531 (1963). TLC techniques.

Cyclopentadienyl Complexes

J. Birmingham in *Advances in Organometallic Chemistry, Vol. 2,* F. G. A. Stone and R. West, Eds., Academic Press, New York, 1964, p. 365. Synthesis of cyclopentadienyl metal compounds.
D. E. Bublitz and K. L. Rinehart, *Organic Reactions, 17,* 1 (1969). The synthesis of substituted ferrocenes and other cyclopentadienyl transition metal complexes.
E. O. Fischer and H. P. Fritz, *Advances in Inorganic Chemistry and Radiochemistry, Vol. 1,* Academic Press, New York, 1959, p. 55. Cyclopentadienyl and other hydrocarbon ring complexes of the transition metals.
H. P. Fritz, *Advances in Organometallic Chemistry, 1,* 240 (1964). Infrared and Raman studies of π-complexes formed between metals and C_nH_n rings.
E. G. Perevalova and T. V. Nikitina, *Organometallic Reactions, 4,* 163 (1972). Reactions of $M(C_5H_5)_2$ complexes.
M. Rosenblum, *Chemistry of the Iron Group Metallocenes: Ferrocene, Ruthenocene, and Osmocene,* John Wiley and Sons, New York, 1965.
G. Wilkinson and F. A. Cotton in *Progress in Inorganic Chemistry, Vol. 1,* F. A. Cotton, Ed., Interscience Publishers, New York, 1959, p. 1. Cyclopentadienyl and arene metal complexes.

$Sn(C_2H_5)_4$, $Sn(C_2H_5)_2Cl_2$, and $Sn(C_2H_5)_2Cl_2 \cdot 2(CH_3)_2SO$

As a member of the group IV elements, tin forms a tetrahedral $SnCl_4$ structure analogous to that of CCl_4. At room temperature both compounds are colorless liquids whose boiling points at atmospheric pressure are 114° and 77°C, respectively. Beyond these comparisons, their chemistries differ considerably. These differences depend to a large extent on the larger size of the Sn atom compared with that of C and on the availability of relatively low energy 5d orbitals on Sn. Both of these factors favor the tendency of Sn to bond to more than four Cl atoms. This is evidenced by the reaction of $SnCl_4$ with Cl^- in water to form the octahedral anion, $SnCl_6^{2-}$.

$$SnCl_4 + 2Cl^- \xrightarrow{H_2O} SnCl_6^{2-}$$

Obviously CCl_4 does not undergo an analogous reaction.

The tendency of $SnCl_4$ to expand its coordination number might also be illustrated by its reactions with numerous donor ligands, L:

$$SnCl_4 + 2L \longrightarrow$$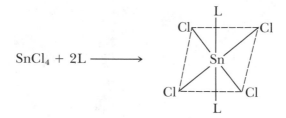

where $L = (CH_3)_3N$, $(C_2H_5)_2O$, or $(C_2H_5)_2S$. While infrared studies suggest that these complexes have a *trans* structure, other L groups are known to give compounds with the *cis* geometry. In moist air, $SnCl_4$ readily fumes to form the compound $SnCl_4 \cdot 5H_2O$, whose structure is

169

unknown but is probably not a simple coordination complex of the preceding type.

In addition to complex formation, $SnCl_4$ forms numerous organotin compounds by reaction with Grignard reagents:

$$SnCl_4 + 4RMgX \rightarrow SnR_4 + 4MgXCl$$

This type of reaction is not limited to $SnCl_4$ but has also been used in the preparation of SiR_4, GeR_4, BR_3, AlR_3, PR_3, AsR_3, and SbR_3 from $SiCl_4$, $GeCl_4$, BCl_3, $AlCl_3$, PCl_3, $AsCl_3$, and $SbCl_3$, respectively. The specific reaction that will be carried out in this experiment is

$$SnCl_4 + 4C_2H_5MgBr \rightarrow Sn(C_2H_5)_4 + 4MgBrCl$$

The product, tetraethyltin, is a colorless liquid that is stable in air and boils at 180°C. Unlike $SnCl_4$, SnR_4 does not form stable 6-coordinated complexes. In general, the tendency to form such complexes decreases as Cl is replaced by R in the compounds $SnCl_yR_{4-y}$: $SnCl_4 >$ $SnCl_3R > SnCl_2R_2 > SnClR_3 > SnR_4$. Because the Sn-C vibrations occur at frequencies below 650 cm^{-1}, the infrared spectrum of $Sn(C_2H_5)_4$ simply consists of C_2H_5 vibrational absorptions that reveal very little about the composition or structure of the compound. Its nmr spectrum is too complex for interpretation without a detailed treatment of nmr spectroscopy. The reason for this complexity is the presence of the isotopes ^{117}Sn and ^{119}Sn (see Appendix 7), which have nuclear spins of ½ and which couple with the ^1H to give a complicated nmr spectrum.

Of somewhat greater value and interest is the mass spectrum of $Sn(C_2H_5)_4$. For a discussion of mass spectrometry, see Experiment 14, p. 140. The large number of stable isotopes of Sn sometimes aids in the assignment of peaks to ion fragments containing Sn, but also can make complex spectra more difficult to interpret. The isotopes of Sn and their natural abundances are:

Isotope	Natural Abundance
^{112}Sn	0.95 per cent
^{114}Sn	0.65 per cent
^{115}Sn	0.34 per cent
^{116}Sn	14.2 per cent
^{117}Sn	7.6 per cent
^{118}Sn	24.0 per cent
^{119}Sn	8.6 per cent
^{120}Sn	33.0 per cent
^{122}Sn	4.7 per cent
^{124}Sn	6.0 per cent

Carbon occurs as 98.89 per cent ^{12}C and 1.11 per cent ^{13}C. The tendency of $Sn(C_2H_5)_4$ to fragment with the loss of ethylene, C_2H_4, and the formation of Sn-H ions should be noted.

In principle, it is possible to synthesize the mixed compounds SnR_yCl_{4-y} by reacting various ratios of $SnCl_4$ and RMgX. In practice, a mixture of compounds that are difficult to separate is obtained. The best method for their preparation is frequently the reaction of stoichiometric amounts of SnR_4 and $SnCl_4$ until exchange of R and Cl has occurred. The equations for these reactions are:

$$3SnR_4 + SnCl_4 \rightarrow 4SnR_3Cl$$

$$SnR_4 + SnCl_4 \rightarrow 2SnR_2Cl_2$$

$$SnR_4 + 3SnCl_4 \rightarrow 4SnRCl_3$$

Yields of SnR_3Cl, SnR_2Cl_2, and $SnRCl_3$ prepared by this method are generally very high. The compound $Sn(C_2H_5)_2Cl_2$ will be synthesized in this experiment from the reaction:

$$Sn(C_2H_5)_4 + SnCl_4 \rightarrow 2Sn(C_2H_5)_2Cl_2$$

The product is a white crystalline solid that melts at 85°. The presence of Cl allows this compound, as contrasted with $Sn(C_2H_5)_4$, to form adduct complexes. Thus, $Sn(C_2H_5)_2Cl_2 \cdot 2(CH_3)_2SO$ (m.p., 64°) precipitates upon mixing ether solutions of $Sn(C_2H_5)_2Cl_2$ and dimethyl sulfoxide (DMSO):

$Sn(C_2H_5)_2Cl_2$ + 2DMSO →

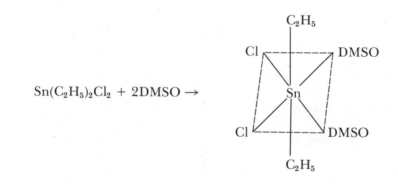

Infrared studies suggest that the product has the geometry shown above. Dimethyl sulfoxide coordinates with certain metals through the S atom, and with others through the O atom. By examining the infrared spectrum, in particular the S-O stretching frequency, the mode of coordination in $Sn(C_2H_5)_2Cl_2 \cdot 2(CH_3)_2SO$ will be established in this experiment.

Not all organotin halides form octahedral complexes. A five-coordinated compound is generated from the reaction of $Sn(CH_3)_3Cl$ with pyridine:

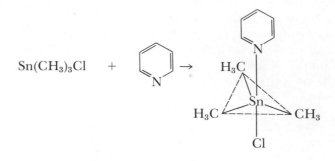

An x-ray diffraction study of the solid indicates that the compound has the trigonal bipyramidal structure shown. Thus tetrahedral, octahedral, and trigonal bipyramidal geometries dominate the structural chemistry of tetravalent tin. The endless variety of organotin complexes that may be prepared makes tin chemistry a fascinating area of research.

EXPERIMENTAL PROCEDURE

Organotin compounds have different toxicities, depending upon the number and type of organo-groups bound to the tin. In general, they are not exceedingly dangerous. If they are handled in a *hood,* they will present no serious hazard. Their relatively low volatilities also minimize any toxic effects that the compounds might have. These preparations should be carried out in a properly functioning hood.

Tetraethyltin, Sn(C₂H₅)₄

As in any Grignard reaction, the reaction vessels should be thoroughly cleaned and oven-dried. This should be done the day before the experiment is actually performed. Put 15.0 g (0.61 mole) of Mg turnings into the 500 ml 3-neck round-bottom flask, and assemble the apparatus as shown in Figure 17–1, which includes a $CaCl_2$ drying tube to protect the reaction mixture from atmospheric moisture. (Grease the ground glass joints, and lubricate the ground glass stirrer bearing with glycerin.)

Add 53 ml (75 g, 0.69 mole) of C_2H_5Br to the addition funnel and 5 ml of diethyl ether to the Mg turnings. Turn on the cool water in the condenser. Then run 5 to 10 ml of the C_2H_5Br onto the Mg turnings. Grignard formation should occur fairly soon, as evidenced by the formation of bubbles on the surface of the Mg. If the reaction does not start, momentarily remove the addition funnel and add one or two drops of

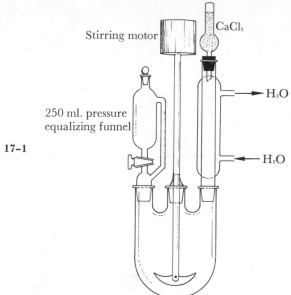

Stirring motor

CaCl₂

H₂O

250 ml. pressure
equalizing funnel

Figure 17-1

H₂O

Br_2 to the solution. The reaction should then start. When the reaction starts, add 45 ml of ether to the flask, stir the Mg turnings, and add the C_2H_5Br dropwise, cooling the flask with an ice bath. This addition will require about 30 minutes. After all of the C_2H_5Br has been added, remove the ice bath and continue stirring until the mixture is at room temperature.

Remove, clean, and dry the addition funnel and add 10 ml (23 g, 0.090 mole) of anhydrous $SnCl_4$ to it. (It is important that the addition funnel be free of ether because of its rapid reaction with $SnCl_4$ to form the solid complex, $SnCl_4 \cdot 2(C_2H_5)_2O$, discussed earlier. It will be noted that the volatility of ether, even at 0°C, is sufficient to cause some of this complex formation in the funnel. The formation of large amounts of the complex will clog the addition funnel.) Cool the reaction flask to 0°C in an ice bath. Add the $SnCl_4$ dropwise to the cooled Grignard solution with continuous stirring over a period of 10 to 15 minutes. This will result in the immediate formation of white $SnCl_4 \cdot 2(C_2H_5)_2O$, and the ether will begin to reflux again. Stir the thick mixture as long as possible. After all of the $SnCl_4$ has been added, remove the ice bath and allow the mixture to come to room temperature.

Then add 50 ml of ether slowly through the addition funnel. Stir the mixture if possible. Replace the drying tube with an adapter (such as that shown in the left neck of the flask in Figure 15-1) and rubber tubing that leads to the back of the hood. *Very cautiously* add to the mixture first 20 ml of ice water and then 100 ml of ice cold 1.2 M HCl. *Rapid addition will produce a violent reaction.* After the addition of some water,

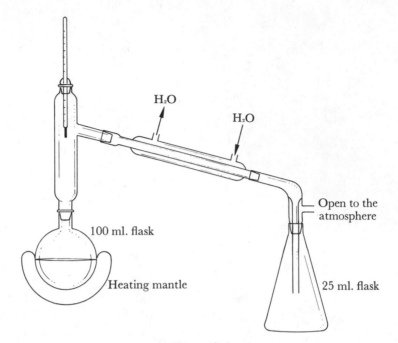

Figure 17–2

start the stirrer as soon as it can be turned by hand. Continue the stirring for 5 minutes after the HCl addition is completed. Separate the ether layer in a separatory funnel and dry it overnight over anhydrous $MgSO_4$.

Suction-filter (Figure 1–1) the solution through a medium glass frit. Transfer the solution to the distillation apparatus illustrated in Figure 17–2, and insulate the distillation head with asbestos tape or glass wool. First distill off the ether and then collect in a 25 ml flask the $Sn(C_2H_5)_4$ boiling at approximately 180° to 182°C at atmospheric pressure. Calculate the percentage yield of the product.

As noted in the earlier discussion, infrared and nmr spectroscopy are not convenient methods of characterizing $Sn(C_2H_5)_4$. The mass spectrum, however, is interesting. If it is not possible to have a mass spectrum run on your sample, analyze the spectrum (Figure E) in the section on NMR and Mass Spectra near the end of the book. Using the isotopic abundances listed earlier, make ion assignments to all the peaks in the spectrum in a manner similar to that outlined in Experiment 14, p. 141. In interpreting the mass spectrum, you should keep in mind that $(C_2H_5)_3SnCl$ is a possible impurity in $Sn(C_2H_5)_4$. [The naturally occurring isotopes of Cl and their abundances are ^{35}Cl (75.4 per cent) and ^{37}Cl (24.6 per cent); see Appendix 7.]

Diethyltin Dichloride, Sn(C₂H₅)₂Cl₂

Place 10 g of $Sn(C_2H_5)_4$ and 11.1 g of $SnCl_4$ (equimolar amounts) in a 100 ml round-bottom flask, and assemble the equipment shown in Figure 17–3. Turn on the cooling water and heat the mixture at 210° to 220°C in a silicone fluid bath for 15 minutes. Raise the flask from the heating bath, and allow it to cool until it can be handled conveniently. Then cool the flask further in a water bath or with running tap water. The product crystallizes as a white solid. Recrystallize the $Sn(C_2H_5)_2Cl_2$ by dissolving it in a minimum amount (~250 ml) of a boiling hydrocarbon solvent with a boiling range of approximately 110° to 140°C. While it is hot, suction-filter the solution (Figure 1–1) and allow it to cool to room temperature. Suction-filter off the white needles of $Sn(C_2H_5)_2Cl_2$, dry, and determine its melting point. Calculate the percentage yield of the product. Measure the infrared spectrum of the product in a Nujol mull (see Experiment 1, p. 19).

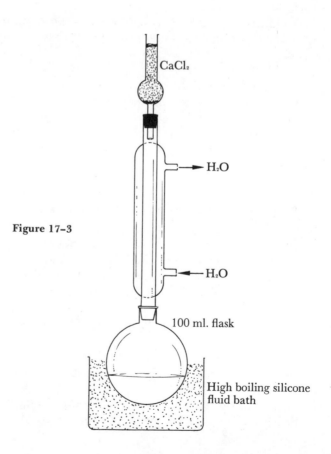

CaCl₂

H₂O

Figure 17–3

H₂O

100 ml. flask

High boiling silicone fluid bath

Sn(C₂H₅)₂Cl₂·2(CH₃)₂SO

Dissolve 5 g (19 mmoles) of $Sn(C_2H_5)_2Cl_2$ in a minimum (~45 ml) of anhydrous ether. Pour this solution into a solution of 3.2 g (41 mmoles) of dimethyl sulfoxide in 5 ml of ether. When the solution is left standing, a white precipitate of $Sn(C_2H_5)_2Cl_2\cdot2DMSO$ separates. Suction-filter, and dry the precipitate on the frit. Calculate the percentage yield and determine the melting point of the product. Record its infrared spectrum in a Nujol mull and compare it with that of $Sn(C_2H_5)_2Cl_2$. The S—O stretching mode of uncoordinated $(CH_3)_2SO$ occurs at approximately 1100 cm^{-1}. Try to locate the S—O absorption in the spectrum of your complex. From the change in frequency from free $(CH_3)_2SO$, suggest whether DMSO is coordinated to Sn through the S or O atom.

REPORT

Include the following:
1. Yield and interpretation of the mass spectrum of $Sn(C_2H_5)_4$. (Make a line drawing spectrum of $Sn(C_2H_5)_4$ that contains only one isotope of Sn.)
2. Yield, melting range, and infrared spectrum of $Sn(C_2H_5)_2Cl_2$.
3. Yield, melting point, and infrared spectrum of $Sn(C_2H_5)_2Cl_2\cdot2DMSO$.
4. Comparison of infrared spectra of $Sn(C_2H_5)_2Cl_2$ and $Sn(C_2H_5)_2Cl_2\cdot2DMSO$.
5. Interpretation of DMSO bonding to Sn in $Sn(C_2H_5)_2Cl_2\cdot2(CH_3)_2SO$.

QUESTIONS

1. Determinations of S or O bonding of $(CH_3)_2SO$ to metals are based on whether the S—O stretching frequency of DMSO in the complex is higher or lower than that of free DMSO. Sulfur bonding is suggested by higher frequencies and oxygen bonding by lower frequencies. What is the basis for interpreting frequencies in this manner?
2. Draw all possible structural and optical (if any) isomers of $Sn(C_2H_5)_2Cl_2\cdot2DMSO$.
3. Suggest methods of analyzing $Sn(C_2H_5)_2Cl_2$ for its percentage Sn and Cl content.
4. Why are H_2O and also HCl added to the reaction mixture during the isolation of $Sn(C_2H_5)_4$? What reaction occurs during the addition of the H_2O?
5. What reactions could be used to prepare $Sn(CH_3)Cl_3$ from $SnCl_4$?
6. Suggest a method for preparing $P(C_2H_5)_3$.

7. The tendency for $Sn(C_2H_5)_yCl_{4-y}$ to coordinate additional ligands decreases as follows:

$$SnCl_4 > Sn(C_2H_5)Cl_3 > Sn(C_2H_5)_2Cl_2 > Sn(C_2H_5)_3Cl > Sn(C_2H_5)_4$$

Explain.

8. Based on reactions that you carried out in this experiment, which ligand do you believe is the stronger Lewis base toward $Sn(C_2H_5)_2Cl_2$ —$(C_2H_5)_2O$ or DMSO?

INDEPENDENT STUDIES

A. Prepare and characterize $Sn(C_2H_5)_3Cl$ or $Sn(C_2H_5)Cl_3$.
B. Prepare and characterize $Sn(C_2H_5)_2F_2$ obtained from the reaction of $Sn(C_2H_5)_2Cl_2$ and aqueous HF. (L. E. Levchuk, J. R. Sams, and F. Aubke, *Inorg. Chem.*, 11, 43 (1972).)
C. Prepare and characterize the pyridine (py) adducts $SnCl_4 \cdot 2py$ and $Sn(C_2H_5)_2Cl_2 \cdot 2py$.
D. Prepare and characterize triphenyltin hydride, $(C_6H_5)_3SnH$. (C. W. Allen, *J. Chem. Educ.*, 47, 479 (1970).)
E. Prepare and characterize $SnCl_2(acetylacetonate)_2$ obtained from $SnCl_4$ and acetylacetone. (D. W. Thompson, D. E. Kranbuehl, and M. D. Schiavelli, *J. Chem. Educ.*, 49, 569 (1972).)

REFERENCES

Sn(C₂H₅)₄, Sn(C₂H₅)₂Cl₂, and Sn(C₂H₅)₂Cl₂·2(CH₃)₂SO

D. B. Chambers, F. Glockling, and M. Weston, *J. Chem. Soc. (A)*, 1759 (1967). Mass spectrum of Sn(C₂H₅)₄ and other organotin compounds.
A. G. Davies, H. J. Milledge, D. C. Puxley, and P. J. Smith, *J. Chem. Soc. (A)*, 2862 (1970). An x-ray structural study of Sn(CH₃)₂Cl₂.
E. Heldt, K. Höppner, and K. H. Krebs, Z. *anorg. allg. Chem.*, 347, 95 (1966). Mass spectra of the compounds Sn(C₂H₅)ᵧ(CH₃)₄₋ᵧ.
N. W. Isaacs and C. H. L. Kennard, *J. Chem. Soc. (A)*, 1257 (1970). An x-ray structural study of Sn(CH₃)₂Cl₂·2(CH₃)₂SO.
K. A. Kozeschkow, *Chem. Ber.*, 66, 1661 (1933). Preparation of Sn(C₂H₅)₂Cl₂.
T. Tanaka, *Inorg. Chim. Acta*, 1, 217 (1967). Preparation and infrared spectra of adducts of SnCl₄ and SnR₂Cl₂.
G. J. M. VanDerKerk and J. G. A. Luijten, *Org. Syntheses*, 36, 86 (1956). Preparation of Sn(C₂H₅)₄.

Organotin Chemistry

J. M. Barnes and L. Magos, *Organometal Chem. Revs.*, 3, 137 (1968), Toxicology of organometallic compounds.
I. R. Beattie, *Quart. Revs.*, 17, 382 (1963). Adducts of quadrivalent Si, Ge, Sn, and Pb compounds.
I. R. Beattie and L. Rule, *J. Chem. Soc.*, 3267 (1964). Infrared studies of SnCl₄ adducts.

G. E. Coates and K. Wade, *Organometallic Compounds, Vol. 1,* Methuen and Co., Ltd., London, 1967, Organometallic chemistry of the nontransition metals.

R. K. Ingham, S. D. Rosenberg, and H. Gilman, *Chem. Revs., 60,* 459 (1960). Comprehensive review of organotin compounds.

K. Nakamoto, *Infrared Spectra of Inorganic and Coordination Compounds,* 2nd Ed., John Wiley and Sons, New York, 1970, p. 210. Infrared spectra of DMSO complexes.

R. Okawara and M. Wada in *Advances in Organometallic Chemistry, Vol. 5,* F. G. A. Stone and R. West, Eds., Academic Press, New York, 1967, p. 137.

R. C. Poller, *The Chemistry of Organotin Compounds,* Academic Press, New York, 1970.

W. L. Reynolds, *Progress in Inorganic Chemistry, 12,* 1 (1970). Dimethyl sulfoxide in inorganic chemistry; primarily metal complexes.

A. K. Sawyer, Ed., *Organotin Compounds,* Marcel Dekker, New York, 1971. A series of volumes.

$K_2S_2O_8$

The industrial preparation of chemicals on a large scale by electrolysis is well known. Aqueous solutions of NaCl are converted to commercially important NaOH, Cl_2, and H_2. Anhydrous molten NaCl yields metallic sodium and Cl_2. Magnesium metal is obtained from $MgCl_2$ by electrolysis, and aluminum is won from Al_2O_3 (Hall process). Metals may be electroplated with Ag, Au, Pt, Cr, Ni, or Cu to give products of greater durability and attractiveness. Aside from commercial applications, electrolysis is sometimes the only method of synthesizing certain chemicals. Fluorine, F_2, is one of these. Since F_2 is the strongest known oxidizing agent, F^- cannot be oxidized to F_2 by any chemical oxidizing agent. Unlimited voltage sources, however, easily oxidize F^-, usually as the fused KHF_2 salt, to F_2. Electrolysis is frequently the synthetic method of choice when the separation of products is difficult. For example, solutions of $CrCl_2$ may be prepared by reducing, with metallic Zn, an aqueous solution of $CrCl_3$:

$$2Cr^{3+} + 6Cl^- + Zn \rightarrow 2Cr^{2+} + 6Cl^- + Zn^{2+}$$

The solution of $CrCl_2$ is, however, contaminated with $ZnCl_2$. To avoid contamination, an aqueous solution of $CrCl_3$ could be electrolyzed to give a pure solution of $CrCl_2$:

$$2CrCl_3 \xrightarrow{\text{elect.}} 2CrCl_2 + Cl_2$$

In this experiment, potassium peroxydisulfate ($K_2S_2O_8$) will be prepared by the electrolysis of an aqueous solution of H_2SO_4 and K_2SO_4.

179

The $S_2O_8{}^{2-}$ ion has been shown by x-ray studies to have the structure,

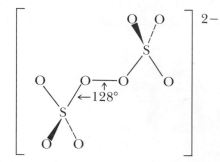

in which the four O atoms around each S atom are tetrahedrally arranged. In the solutions to be electrolyzed, K^+ and $HSO_4{}^-$ are the major species present. An electrical current is passed through the solution, as schematically shown in Figure 18–1.

The reaction at the cathode is

$$2H^+ + 2e^- \rightarrow H_2$$

The desired reaction at the anode is the oxidation:

$$2HSO_4{}^- \rightarrow S_2O_8{}^{2-} + 2H^+ + 2e^-, \qquad E° = -2.05 \text{ volts}$$

It is obvious, however, that the oxidation of H_2O to O_2,

$$2H_2O \rightarrow O_2 + 4H^+ + 4e^-, \qquad E° = -1.23 \text{ volts}$$

has a higher oxidation potential and should occur in preference to that of $HSO_4{}^-$ oxidation. The value of $E° = -1.23$ volts for the oxygen

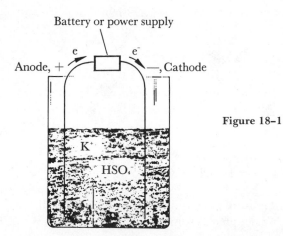

Figure 18–1

half-reaction is not obtained from electrochemical measurements, but rather from other thermodynamic sources that allow the calculation of E° from $\Delta H°$, $\Delta S°$, and $\Delta G°$ values.

In practice, much more than -1.23 volts is required to liberate O_2 from water. This results from the irreversibility of this half-reaction, which is caused by a slow step in the oxidation of H_2O. The kinetically slow step requires additional voltage ("overvoltage") before O_2 is formed. The rate of this slow reaction, whatever it might be, is greatly affected by the composition of the electrode at which the oxidation occurs. Thus, the overvoltage for the O_2 electrode in 1 M KOH varies with the anode material as follows (current density = 1.0 amp/cm²):

Anode	Overvoltage
Ni	0.87 volts
Cu	0.84
Ag	1.14
Pt	1.38

These overvoltages are not very reproducible and depend upon the history of the anode material, but their differences suggest that the electrode participates in the slow step of the oxidation process. To be sure, overvoltages are a familiar but poorly understood phenomenon. For the purposes of a synthetic chemist, the oxygen overvoltage permits the oxidation of substances in H_2O, which would not be possible if the $H_2O \rightarrow O_2$ couple exhibited no overvoltage. Because of the high oxygen overvoltage observed for Pt, this will be used as the anode material in the preparation of $K_2S_2O_8$.

To maximize the yield of $K_2S_2O_8$ and minimize O_2 formation, it is advantageous to adjust other conditions of the electrolysis to increase the oxygen overvoltage. Since overvoltages increase with current density (*vide infra*), relatively high currents will be used. Also, if the electrolysis is carried out at low temperature, reaction rates decrease, and the rate of the slow step in the oxidation of H_2O will decrease. This will increase the oxygen overvoltage. Low temperatures should therefore favor the formation of $S_2O_8^{2-}$. Finally, high concentrations of HSO_4^- and low H_2O concentrations maximize $K_2S_2O_8$ yields. For these reasons, the electrolysis of HSO_4^- to form $S_2O_8^{2-}$ will be carried out using (1) a Pt electrode, (2) high current density, (3) low temperature, and (4) a saturated solution of HSO_4^-. Careful consideration of factors such as those that have been mentioned has allowed the commercial electrolytic preparation and purification of chemicals on a large scale.

As in any electrolytic preparation, there is always some danger of the product, which is generated at the anode, diffusing to the cathode and undergoing reduction to the starting material. Generally the cathode and anode compartments must be separated and connected by a bridge to prevent this from occurring. In this experiment, the $S_2O_8^{2-}$ that is produced at the anode would be expected to diffuse toward the cathode, where it would be the most readily reduced species in solution; it would be immediately reduced to HSO_4^-. Fortunately, $K_2S_2O_8$ is quite insoluble in water, and it precipitates from solution before it reaches the cathode.

The Pt anode will be relatively small diameter wire (22 gauge). Knowing the diameter of the anode wire (22 gauge B and S wire has a diameter of 0.0644 cm) and the length of it coming in contact with the HSO_4^- solution, it is possible to calculate the current density, which is defined as

$$\text{current density} = \frac{\text{amperes}}{\text{area of anode}}$$

The desired current density for this synthesis is 1.0 amp/cm². The number of amperes to be passed through the anode should be sufficient to give this current density.

The amount of product generated will, of course, depend on the electrical current (i.e., the number of electrons) passed through the solution. Since the product of the current, I, in amperes, and the time, t, in seconds, of current passage gives coulombs of electricity, and 96,500 coulombs oxidize (or reduce) one equivalent of reactant, the theoretical yield of product will be

$$\text{theoretical yield} = \frac{\text{coulombs passed}}{96,500 \text{ coulombs/equiv.}} \times (\text{equiv. wt.}) =$$

$$\frac{(I)(t)}{96,500}(\text{equiv. wt.})$$

Since the actual yield is frequently less than this number of grams because of side reactions, percentage yields are usually evaluated. In electrochemistry percentage yield is called current efficiency:

$$\text{percentage yield} = \text{current efficiency} = \frac{\text{actual yield}}{\text{theoretical yield}} \times 100$$

Salts of the peroxydisulfate ion, $S_2O_8^{2-}$, are relatively stable but in acidic solution react to give H_2O_2:

$$O_3S-O-O-SO_3^{2-} + 2H^+ \rightarrow HO_3S-O-O-SO_3H$$

$$HO_3S-O-O-SO_3H + H_2O \rightarrow HO_3S-O-OH + H_2SO_4$$

$$HO_3S-O-OH + H_2O \rightarrow H_2SO_4 + H_2O_2$$

Under certain conditions, it is possible to stop the reaction at the intermediate peroxymonosulfuric acid, HO_3SOOH, but in the commercial preparation of H$_2$O$_2$ the reactions are driven to completion by distilling off the hydrogen peroxide.

The S$_2$O$_8{}^{2-}$ ion is one of the strongest known oxidizing agents and is even stronger than H$_2$O$_2$.

$$S_2O_8{}^{2-} + 2H^+ + 2e^- \rightarrow 2HSO_4{}^-, \qquad E° = 2.05 \text{ volts}$$

$$H_2O_2 + 2H^+ + 2e^- \rightarrow 2H_2O, \qquad E° = 1.77 \text{ volts}$$

It will oxidize many elements to their highest oxidation states, as will be illustrated by the reactions that you will carry out with K$_2$S$_2$O$_8$. For example, Cr^{3+} may be oxidized to Cr$_2$O$_7{}^{2-}$ according to the equation:

$$3S_2O_8{}^{2-} + 2Cr^{3+} + 7H_2O \rightarrow 6SO_4{}^{2-} + Cr_2O_7{}^{2-} + 14H^+$$

Like most S$_2$O$_8{}^{2-}$ oxidations, the reaction is relatively slow, but is catalyzed by adding Ag$^+$ ion. A kinetic study of the Ag$^+$ catalyzed reaction suggests that the initial step is the oxidation of Ag$^+$ to the very reactive Ag^{3+}:

$$S_2O_8{}^{2-} + Ag^+ \rightarrow 2SO_4{}^{2-} + Ag^{3+}$$

This Ag^{3+} then rapidly reacts with Cr^{3+} to give Cr$_2$O$_7{}^{2-}$ and regenerate the Ag$^+$:

$$3Ag^{3+} + 2Cr^{3+} + 7H_2O \rightarrow Cr_2O_7{}^{2-} + 3Ag^+ + 14H^+$$

The details of these reactions are unknown, but S$_2$O$_8{}^{2-}$ oxidations frequently proceed by an initial O—O bond breaking in S$_2$O$_8{}^{2-}$ to give the highly reactive radical anions SO$_4{}^-$, which carry out the further oxidations.

The strong oxidizing powers of S$_2$O$_8{}^{2-}$ have also permitted the syntheses of coordination complexes of silver in the unusual oxidation state of 2+. The synthesis to be carried out is that of the complex [Ag(pyridine)$_4$]S$_2$O$_8$:

$$2Ag^+ + 3S_2O_8{}^{2-} + 8py \rightarrow 2[Ag(py)_4]S_2O_8 + 2SO_4{}^{2-}$$

The cation, Ag(py)$_4{}^{2+}$, has a square planar geometry, analogous to that of Cu(py)$_4{}^{2+}$.

EXPERIMENTAL PROCEDURE

Electrolysis Cell

Cell construction is very simple. The anode is made by sealing a 22 gauge platinum wire into 6 mm glass tubing. The length of anode that

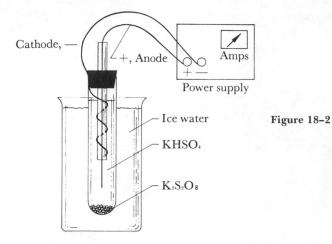

Cathode, —

+, Anode

Amps

+ —

Power supply

— Ice water

— KHSO₄

— K₂S₂O₈

Figure 18–2

is in contact with the solution is about 6 cm. The cathode is a Pt wire wound around the glass tubing (Figure 18–2). Insert the electrode assembly into a cork or rubber stopper that either contains a hole or is loosely fitted into the approximately 2×20 cm test tube. These measures allow gaseous reaction products to escape from the system. An adjustable power supply conveniently provides the 1.0 amp/cm² current density required for the $K_2S_2O_8$ preparation. Note that this amperage level is dangerous, and all electrode connections should be made with care.

Potassium Peroxydisulfate, $K_2S_2O_8$

About 3 g of $K_2S_2O_8$ is required for subsequent reactions. Prepare a saturated solution of $KHSO_4$ by saturating a solution of 150 ml of water and 60 ml of concentrated H_2SO_4 with K_2SO_4. This is probably best done by adding K_2SO_4 to the heated H_2SO_4 solution until no more K_2SO_4 dissolves. Then cool the solution to 0°C in an ice bath overnight to ensure that precipitation of excess K_2SO_4 is complete. Pour the supernatant solution into the electrolysis cell and immerse the cell in an ice bath (Figure 18–2). Turn on the power supply (record the time and amperage) and adjust the amperage until the anode current density is 1 amp/cm². The amperage required will be determined by the area of the anode, as noted in the earlier discussion. Use the amperage that is required by the area of your anode to achieve a 1 amp/cm² current density. Allow the current to flow for 30 to 45 minutes, during which time white crystals of $K_2S_2O_8$ collect on the bottom of the tube. The reaction will slow considerably toward the end of this period owing to depletion of HSO_4^-. The resistance of the solution to the current will generate sufficient heat to require replenishing the ice in the bath during electrolysis.

After the reaction period, turn off the power supply and record the time. Suction-filter (Figure 1–1) the K$_2$S$_2$O$_8$ crystals and wash them on the frit, first with 95 per cent ethanol and finally with ethyl ether. Determine the yield. From the amperage and time, calculate the current efficiency. If 3 g of K$_2$S$_2$O$_8$ is not obtained, add fresh HSO$_4$$^-$ solution to the cell and repeat the synthesis.

Reactions of K$_2$S$_2$O$_8$. Prepare a saturated solution of K$_2$S$_2$O$_8$ by dissolving about 0.75 g of K$_2$S$_2$O$_8$ in a minimum of water. React this K$_2$S$_2$O$_8$ solution with each of the solutions listed below. What happens in each case? (Use test tubes as reaction vessels.)

1. Reaction with an acidified KI solution. Warm slightly.
2. Reaction with MnSO$_4$ in dilute H$_2$SO$_4$ to which 1 drop of AgNO$_3$ solution is added. Warm gently.
3. Reaction with Cr$_2$(SO$_4$)$_3$ in dilute H$_2$SO$_4$ to which 1 drop of AgNO$_3$ solution is added. Warm gently.
4. Reaction with AgNO$_3$ solution.

For comparison with the K$_2$S$_2$O$_8$ reactions, react each of the above solutions with 30 per cent H$_2$O$_2$. What happens in each case?

Tetrapyridinesilver(II) Peroxydisulfate, [Ag(py)$_4$]S$_2$O$_8$

Add 1.4 ml of a good grade of pyridine to 3.2 ml of an aqueous solution containing 0.16 g of AgNO$_3$. With stirring, pour this solution into a solution of 2.0 g of K$_2$S$_2$O$_8$ dissolved in 135 ml of water. After the solution has been left standing for 30 minutes, suction-filter the solution (Figure 1–1). Wash the yellow product with a small amount of water and dry in a desiccator. Calculate the percentage yield and record the infrared spectrum of a Nujol mull of [Ag(py)$_4$]S$_2$O$_8$ (Experiment 1, p. 20).

REPORT

Include the following:
1. Calculation of the amperage to be used to attain a current density of 1 amp/cm^2.
2. Current efficiency in K$_2$S$_2$O$_8$ preparation.
3. Write balanced equations for the reactions of K$_2$S$_2$O$_8$ and H$_2$O$_2$ with KI, MnSO$_4$, Cr$_2$(SO$_4$)$_3$, and AgNO$_3$.
4. Percentage yield of [Ag(py)$_4$]S$_2$O$_8$ and assignments of infrared absorption bands to vibrations in the compound.

QUESTIONS

1. Give at least two reasons for the low current efficiency in the preparation of K$_2$S$_2$O$_8$. How would you experimentally determine which

of these factors was most important in reducing the current efficiency?

2. Draw a Lewis diagram (dot formula) for the $S_2O_8^{2-}$ ion.
3. The $Ag(py)_2^+$ complex is colorless, while $Ag(py)_4^{2+}$ is colored. Explain why this is expected.
4. From the standard oxidation potentials of $S_2O_8^{2-}$ and H_2O, would you expect $S_2O_8^{2-}$ to oxidize H_2O to O_2 and H^+? In fact, does this reaction occur? Why or why not?
5. Give the formula of at least one other compound, besides $[Ag(py)_4]$-S_2O_8, that contains silver in the 2+ oxidation state.
6. The $K_2S_2O_8$ that is produced may be contaminated with K_2SO_4 or $KHSO_4$. Suggest a method for determining the purity of the $K_2S_2O_8$ product.
7. Write an equation for the overall reaction that occurs during the electrolysis of the aqueous $KHSO_4$ solution.
8. Why can K_2SO_4 not be used in place of $KHSO_4$ in the preparation of $K_2S_2O_8$?
9. Why is it important that the cathode and anode not be too close to each other in the electrolysis solution?
10. If, instead of Pt, a Cu wire were used as the anode, would $K_2S_2O_8$ still be formed? Explain.

INDEPENDENT STUDIES

A. Analyze your $K_2S_2O_8$ product to determine its purity.
B. Measure the infrared spectrum of $K_2S_2O_8$ and make assignments to as many absorptions as possible.
C. Prepare $[Ag(dipyridyl)_2](NO_3)_2$ by electrolysis of a solution of $AgNO_3$ and dipyridyl. (W. G. Thorpe and J. K. Kochi, *J. Inorg. Nucl. Chem., 33*, 3958 (1971).)
D. Prepare AgO by electrolysis of $AgNO_3$. (W. L. Jolly, *Synthetic Inorganic Chemistry*, Prentice-Hall, Englewood Cliffs, N.J., 1960, p. 148. W. L. Jolly, *The Synthesis and Characterization of Inorganic Compounds*, Prentice-Hall, Englewood Cliffs, N.J., 1970, p. 448.)
E. Prepare amalgams of Ba or the lanthanides by electrolysis of metal salt solutions, using a Hg cathode. (B. C. Marklein, D. H. West, and L. F. Audrieth, *Inorganic Syntheses, 1*, 11 (1939). E. E. Jukkola, L. F. Audrieth, and B. S. Hopkins, *Inorganic Syntheses, 1*, 15 (1939).)
F. Prepare $KClO_3$ by electrolysis of KCl. (G. Pass and H. Sutcliffe, *Practical Inorganic Chemistry*, Chapman and Hall Ltd., London, 1968, p. 84. H. F. Walton, *Inorganic Preparations*, Prentice-Hall, Englewood Cliffs, N.J., 1948, p. 169.)

REFERENCES

$K_2S_2O_8$

D. M. Adams and J. B. Raynor, *Advanced Practical Inorganic Chemistry,* John Wiley and Sons Ltd., London, 1965, p. 122.
D. A. House, *Chem. Rev., 62,* 185 (1962). Kinetic studies of oxidations by peroxydisulfate.
W. L. Jolly, *Synthetic Inorganic Chemistry,* Prentice-Hall, Englewood Cliffs, New Jersey, 1960, pp. 74–85, 143.
H. F. Walton, *Inorganic Preparations,* Prentice-Hall, Englewood Cliffs, New Jersey, 1948, pp. 59–67, 167.
D. M. Yost, *J. Am. Chem. Soc., 48,* 152 (1926). Catalysis of $Cr^{3+} \rightarrow Cr_2O_7^{2-}$ by Ag^+.

Electrolytic Synthesis

J. Chang, R. F. Large, and G. Popp, in *Physical Methods of Chemistry, Vol. I, Part IIB,* A. Weissberger and B. Rossiter, Eds., John Wiley and Sons, Inc., 1971. Brief survey of electrochemical syntheses and experimental techniques.
A. M. Feltham and M. Spiro, *Chem. Rev., 71,* 177 (1971). Factors that affect the behavior of platinized platinum electrodes.
J. B. Headridge, *Electrochemical Techniques for Inorganic Chemists,* Academic Press, New York, 1969. A very brief, general discussion of electrolysis in inorganic synthesis (pp. 57–61).
B. L. Laube and C. D. Schmulbach, *Progress in Inorganic Chemistry, 14,* 65 (1971). Electrosynthesis of inorganic compounds in non-aqueous solvents.
See references above by Walton and by Jolly.

Redox Reactions

F. Basolo and R. G. Pearson, *Mechanisms of Inorganic Reactions,* 2nd Ed., John Wiley and Sons, 1967, Chapter 6. Oxidation-reduction reactions of coordination compounds.
S. B. Brown, P. Jones, and A. Suggett, *Progress in Inorganic Chemistry, 13,* 159 (1970). Kinetic studies of redox reactions of peroxides.
W. M. Latimer, *Oxidation Potentials,* 2nd Ed., Prentice-Hall, Englewood Cliffs, New Jersey, 1952. Redox chemistry of the elements.
H. Taube, *J. Chem. Educ., 45,* 452 (1968). Mechanisms of oxidation-reduction reactions of primarily transition metal complexes.

$(CH_3)_3N:BF_3$

Note: A continuous period of 4 to 5 hours is generally required to complete the work on the vacuum line.

While the use of simple high vacuum systems in synthetic inorganic chemistry dates back to the origins of chemistry, the technique as it is known today was essentially established by Alfred Stock and his co-workers in their pioneering research on the boron hydrides in the 1920's. The volatility and extreme reactivity of these hydrides (B_2H_6 spontaneously burns in air) made them particularly suitable for study in a high vacuum system. Although high vacuum techniques may be very highly specialized and sophisticated, the purpose of this experiment is to introduce only a few of those techniques that are the most fundamental and important. They are transferring chemicals from one container to another on the vacuum line, measuring the vapor pressure of a liquid at low temperature, and determining the molecular weight of a gas.

As the name implies, vacuum lines are usually operated under relatively low pressures—substantially below atmospheric pressure. The unit used for expressing gaseous pressures is the *torr,* and one torr is equal to 1 mm Hg pressure. The mechanical oil pump that will be used in this experiment to create the vacuum should be capable of reducing the pressure within the glass apparatus to 10^{-3} or 10^{-4} torr. (It should be remembered that the vapor pressure of Hg is 1.7×10^{-3} torr at 24°C, and any quoted pressures exclude any Hg vapor present in the apparatus. A pressure of 10^{-4} torr means that the pressure exerted by other gases besides Hg is 10^{-4} torr.) Low pressures accomplish two objectives. They remove any undesired reactive gases such as O_2 or H_2O, and they allow volatile substances to be rapidly transported from one portion of the apparatus to another.

To illustrate the latter point, consider the distillation or transfer of a liquid from trap A to B in Figure 19–1. In order to make the transfer, the temperature of B must be lower than that of A. Therefore, one

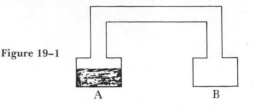

Figure 19–1

A   B

might immerse trap B in a Dewar flask containing liquid nitrogen (b.p. $-196°C$). (See Appendix 9 for other cold baths.) The gaseous molecules in B will condense at this temperature. Liquid molecules in A vaporize and diffuse by virtue of their kinetic energy to B and are condensed. The rate of diffusion of the molecules to B depends upon the number of collisions that they encounter on the way to B. If the system is filled with an inert gas such as He, the diffusion of molecules from A to B is very slow because of the large number of collisions that the diffusing molecules have with He atoms. On the other hand, if the He pressure is reduced to 10^{-3} or 10^{-4} mm Hg, there are many fewer molecules to get in the way, and distillation occurs more rapidly. In vacuum it is possible to make such transfers with any volatile compound. The volatility is, of course, determined by the temperature of the liquid; compounds having boiling points of $150°C$ at atmospheric pressure may be manipulated in a vacuum system without difficulty.

The vacuum line to be used in this experiment is shown in Figure 19–4. It is attached to a mechanical, rotary vacuum pump. The oil in the pump should be of good grade and changed regularly. After extended use the oil accumulates chemicals handled in the vacuum line. They will increase the vapor pressure of the oil and decrease the efficiency of the pump. Since many of these chemicals are toxic, a tube should be attached to the exhaust port of the pump and vented into a hood. To protect the pump from such gases, it is important that a large-volume cold trap be placed between the pump and the working portion of the vacuum line (Figure 19–4). The cold trap is cooled with a Dewar flask containing liquid N_2. While liquid N_2 is sufficiently cold ($-196°C$) to trap most gases used in the line, it also is cool enough to condense liquid O_2 (b.p. $-183°C$) into the trap if the trap is open to air. Because liquid O_2 is such a powerful oxidizing agent, it may react explosively with any other gases that may have also condensed into the trap. For this reason, *the cold trap should never be opened to the air until the liquid N_2 trap is removed.*

A vacuum line is frequently made up of several sections joined together with ground glass joints (Figure 19–4). These joints may be either of the usual standard taper type or ball and socket joints (Figure 19–2); both types are used in the apparatus in Figure 19–4. The standard taper joint is usually less likely to leak and is used where bending

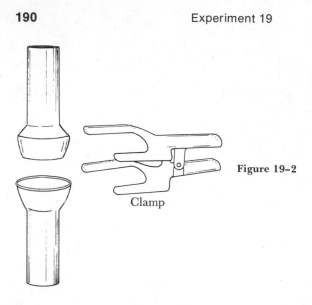

Figure 19–2

Clamp

at the joint is not required. Where some flexibility is needed, the ball and socket joint should be used. Both joints should be greased with either a high vacuum silicone or Kel-F grease. The latter grease will be used in this experiment because BF_3 reacts with silicone greases.

High vacuum stopcocks (Figure 19–3), which may be evacuated and do not pop out even when the pressure in the line is somewhat above atmospheric pressure, should be greased before being used. An ungreased stopcock should never be turned; it will become scored. Since the ground glass barrel and key in a vacuum stopcock are carefully matched, it is important that these always be kept together; a barrel without its *matching* key is worthless. Since stopcocks will break off the apparatus when put under strain and are difficult and expensive to replace, they should be turned with both hands, one to support the barrel and the other to turn the key. It is important to realize the cost of carelessness when working with a glass vacuum system.

Although a thorough understanding of the vacuum line and its use is the most important aspect of this experiment, a few comments should be made about the reaction that will be carried out in the last

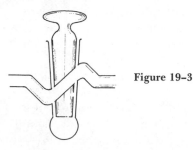

Figure 19–3

phase of the experiment. It is a simple reaction between a gaseous Lewis acid, BF$_3$, and a gaseous Lewis base, (CH$_3$)$_3$N, to form the solid adduct (CH$_3$)$_3$N:BF$_3$.

$$BF_3 + (CH_3)_3N \rightarrow (CH_3)_3N:BF_3$$

The product, (CH$_3$)$_3$N:BF$_3$, is a white solid that is stable in air and has a melting point of 146°C. It shows little tendency to dissociate and may even be sublimed under vacuum. The reactant, BF$_3$, is a colorless gas that fumes in moist air. It has a boiling point of −101°C at 760 torr pressure and a melting point of −127°C. Its low boiling point means that solid or liquid BF$_3$ may be easily and inadvertently converted to a gas. If this occurs when the BF$_3$ is confined to a small portion of the vacuum line, the pressure created by the gaseous BF$_3$ will easily shatter the glass vacuum system. (If the vacuum line is designed so that it is possible to confine the gas in a small volume of the line, a safety shield placed between the worker and the line is strongly advised.) The other reactant, (CH$_3$)$_3$N, melts at −117°C and boils at 2.9°C (at 760 torr).

The simplicity of this type of reaction has led numerous researchers to investigate the relative acidities and basicities of numerous acids, A, and bases, B. These parameters have been established by comparing equilibrium constants for the reaction

$$A + B: \overset{K}{\rightleftharpoons} B:A$$

using different acids and bases. In some cases, it has been possible to measure ΔH's for the reactions. For example, enthalpies for the exothermic reaction of pyridine (C$_5$H$_5$N) with the boron trihalides, BX$_3$,

$$C_5H_5N + BX_3 \rightarrow C_5H_5N:BX_3$$

decrease with BX$_3$ in the order: BBr$_3$ > BCl$_3$ > BF$_3$. Thus, BBr$_3$ forms a more stable adduct with pyridine than does BF$_3$. Electronegativity arguments suggest that BF$_3$ would be a stronger Lewis acid than BBr$_3$. On closer consideration of the reaction, one notes that a considerable reorganization of bonds occurs around boron in going from the reactant to the product. The boron trihalides are planar molecules, whereas the geometry of the adducts is essentially tetrahedral.

Thus, a rehybridization of boron occurs during the reaction, and the stability of the planar BX_3 relative to the adduct is a major factor determining the ΔH of the reaction. The planarity and unusually short B-X bond lengths in BX_3 are attributed to π bonding from the $p\pi$ orbitals on the halogens, X, to the empty $p\pi$ orbital on boron. Since the $p\pi$ orbitals of F are smaller and less diffuse than those of Br, F-B $p\pi$-$p\pi$ bonding is greater than is π-bonding in the analogous Br-B bonds. For this reason, BF_3 is relatively more stable and has a lesser tendency than BBr_3 to form an adduct complex. Numerous other thermodynamic studies of adduct formation by various Lewis acids and bases have led to an understanding of many of the factors that determine the stabilities of these complexes.

EXPERIMENTAL PROCEDURE

BF_3 should not be exposed to air at any time during the experiment, since it fumes and is toxic. A more serious danger is one already mentioned. Namely, at temperatures above $-101°C$, BF_3 is a gas; if it is allowed to vaporize in a closed section of the vacuum line, the BF_3 pressure will shatter the vacuum line and persons or equipment nearby. The vacuum system shown in Figure 19–4 has relatively few locations in which the BF_3 may be closed off from a mercury manometer. If BF_3 vaporizes, the mercury will be forced down in the manometer and may be completely displaced from the manometer tube, thus preventing an explosion.

Prior to starting this experiment, read the experiment very carefully, and if any operation is not clear, ask for help. Although a vacuum line can be dangerous if improperly used, no difficulties should be encountered if the instructions are followed carefully. Take your time!

In the first portion of the experiment you will introduce BF_3 into the line, transfer it from trap to trap, measure its vapor pressure at $-111.6°C$, determine its molecular weight, and finally react it with $(CH_3)_3N$. The product, $(CH_3)_3N:BF_3$, will then be sublimed under vacuum and characterized by its melting point and infrared spectrum.

If the vacuum line is already assembled, turn (use both hands!) all of the stopcocks to determine whether they open and close easily. Regrease with Kel-F vacuum grease any that do not. If the vacuum line is not complete, clamp the necessary equipment on the supporting rack in the order shown in Figure 19–4. Place the clamps near the stopcocks where stress will be greatest during use. Make certain that the clamps offer firm support of the glassware *but do not introduce severe strains where breakage might occur.* Be sure to clamp the ball and socket joints (Figure 19–2). On joint 1 (J1), clamp the 200 ml bulb to be used in the molecular weight determination. Close the inlet stopcock and

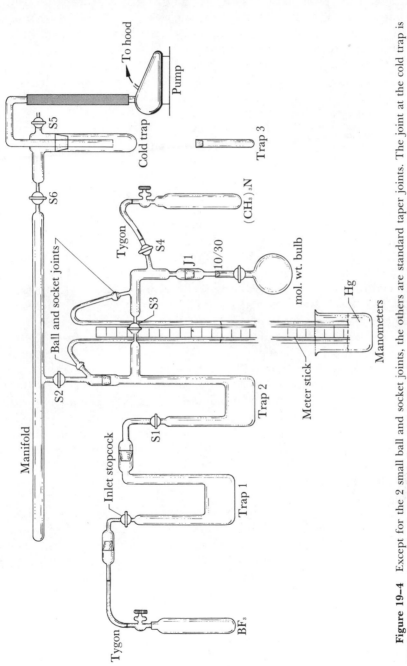

Figure 19–4 Except for the 2 small ball and socket joints, the others are standard taper joints. The joint at the cold trap is 45/50 and that at the molecular weight bulb is 10/30. The other standard taper joints are 24/40.

stopcocks S4 and S5; open all of the others. After greasing the male joint of the cold trap, twist on the trap to give a tight seal. Start the pump and evacuate. Turn all of the stopcocks so as to evacuate the stopcock bulbs. When the line is pumped down (listen to the pump and look at the manometers, which should read atmospheric pressure, $\sim$740 to 750 mm Hg), place a Dewar flask of liquid N_2 ($-196°C$) around the cold trap. To minimize evaporation, cover the open part of the Dewar with a cloth towel. Do not allow the level of liquid N_2 to rise above the bottom of the standard taper joint of the trap.

The order of the preceding operations is important, for if the trap is cooled before evacuating the line, liquid O_2 (b.p., $-183°C$) will condense in the cold trap. This will result in a poor vacuum, since liquid O_2 has an appreciable vapor pressure at $-196°C$. Also, an explosive mixture might result if organic vapors were later condensed into the cold trap during the experiment.

At this point the inlet stopcock and S4 and S5 should be closed, while all other stopcocks are open. Close stopcock S3. With Tygon tubing, connect Trap 1 to the BF_3 compressed gas lecture bottle cylinder. With a wrench, open the main valve on the BF_3 lecture bottle cylinder; make certain that the needle valve on the cylinder is closed. Open the inlet stopcock slowly to evacuate the Tygon tubing. Close S2, thus separating the manifold and pump from the traps.

Next, place a Dewar flask containing liquid N_2 around Trap 1. Cover just the bottom portion of the trap with liquid N_2; if the arms of the trap are cooled, the solid will condense in them, reducing the capacity of the trap. You will now be able to monitor the BF_3 pressure in Trap 1 by observing the mercury manometer on the left. Open the needle valve on the lecture bottle slowly, noting the pressure. Deposit 2 to 3 g (estimate) of BF_3 (density of solid is 1.9 g/ml) in Trap 1. Do not allow the pressure to rise above 200 torr. When sufficient BF_3 has been added, close the needle valve on the lecture bottle, close the inlet stopcock, and finally close the main valve on the lecture bottle. Any leaks in the system will be apparent, as BF_3 fumes in air.

Prepare a liquid N_2-CS_2 slush bath in a small Dewar flask. Fill the Dewar to within 8 cm of the top with CS_2. While stirring vigorously, add liquid N_2 until the slush has the consistency of a thick malted milk. (Do not allow a crust to form on top or the Dewar could break.) The temperature of the slush bath is the melting point of CS_2, $-111.6°C$.

Remove the liquid N_2 Dewar from Trap 1 and put it around Trap 2. Allow the BF_3 to liquefy. Control the rate of warming with the CS_2 slush, but do not allow the pressure to exceed 280 to 290 torr. When a pressure of approximately 280 torr is observed, place the CS_2 slush bath around Trap 1. Distill all of the BF_3 into Trap 2. When the transfer is complete as indicated by the pressure, close S1.

Open S2, S3, and the stopcock on the molecular weight bulb, and equalize the pressure in the two Hg columns. Close S2 and S3. Allow

Trap 2 to warm to $-111.6°C$ and place the CS_2 slush bath around it. Measure the vapor pressure of BF_3 at this temperature. It should be roughly 300 torr.

Open S3, allowing the preweighed and evacuated molecular weight bulb to be filled with BF_3 at the observed pressure. Close the stopcock on the bulb and condense the residual BF_3 back into Trap 2 with the liquid N_2 Dewar. Close S3 and open S4, allowing the removal of the bulb. (Note that the $(CH_3)_3N$ lecture bottle in Figure 19–4 has not yet been attached.) After wiping off the stopcock grease, weigh the bulb; from its volume (fill with water and weigh; determine the volume using the density of water), temperature, and pressure, calculate the molecular weight of the gas.

Attach Trap 3 to joint 1 (J1) and join the $(CH_3)_3N$ compressed gas lecture bottle cylinder to S4 with Tygon tubing. Evacuate the system by opening S2, S3, and S4. Then close S2 and S3. Replace the liquid N_2 Dewar on Trap 2 with the CS_2 slush. Place the liquid N_2 Dewar around Trap 3. Open the main valve on the $(CH_3)_3N$ lecture bottle, and then with the needle valve, allow about 5 g of $(CH_3)_3N$ to slowly condense into Trap 3. Close the needle valve, S4, and the main valve on the cylinder. Then *slowly* condense *all* of the BF_3 in Trap 2 into Trap 3 by opening S3. The heat generated by this highly exothermic reaction will significantly increase the pressure in the system, and if uncontrolled will force the mercury out of the manometer. The BF_3 should therefore be added slowly and in about three portions. After adding each portion, remove the N_2 Dewar and allow the mixture to liquefy and react. When the BF_3 is completely transferred, replace the liquid N_2 Dewar with the CS_2 slush and allow Trap 3 to warm to $-111.6°C$ over a 15 minute period. Any excess gas remaining after reaction is pumped into the cold trap by opening stopcock S2. The product, $(CH_3)_3N:BF_3$, in Trap 3 is freed of excess gases by allowing it to remain in the dynamic vacuum for 30 minutes at room temperature. Close S3, remove the $(CH_3)_3N$ lecture bottle, open S4, and remove Trap 3. The product is stable in air.

Shut down the vacuum line as follows: (1) remove the liquid N_2 Dewar from the cold trap; (2) turn off the vacuum pump; (3) open S5; and (4) remove the cold trap and place it in the hood, to allow volatile substances to evaporate.

Sublime $(CH_3)_3N:BF_3$ under vacuum at $110°C$ (Figure 11–3). Determine the melting point of the sublimed product in a sealed (at both ends) capillary tube. Record the infrared spectrum of the adduct in $CHCl_3$ or CH_2Cl_2 solvent (Experiment 14, p. 136).

REPORT

Include the following:
1. Measured vapor pressure of BF_3 at $-111.6°C$.

2. Data and calculations for the determination of the molecular weight of BF_3. Possible errors.
3. Melting point of $(CH_3)_3N:BF_3$.
4. Infrared spectrum and band assignments to vibrations in $(CH_3)_3N: BF_3$.

QUESTIONS

1. Write a balanced equation for the fuming of BF_3 in air. How serious a safety hazard are the products?
2. If 2 g of BF_3 were condensed into Trap 1 at $-196°C$ and both the inlet stopcock and S1 were closed, estimate the pressure (in atmospheres) that would develop if the trap were allowed to warm up to room temperature.
3. You want to determine the percentage yield of the reaction between a known amount of $(CH_3)_3N$ and a large excess of BF_3. How would you modify the experiment so that you would know the amount of $(CH_3)_3N$ undergoing reaction?
4. In making a liquid N_2-slush bath that had a temperature of $-100°$, how would you proceed in selecting a liquid that, when mixed with liquid N_2, would give you that temperature?
5. At high temperature $(CH_3)_3N:BF_3$ vaporizes to give a gas. How would you determine experimentally whether the gas was $(CH_3)_3N:BF_3$ or a mixture of $(CH_3)_3N$ and BF_3?
6. The proton nmr spectrum of $(CH_3)_3N:BCl_3$ consists of four peaks of equal intensity, which are separated by 2.6 Hertz (cycles per second). In contrast, the spectrum of $(CH_3)_3N:BF_3$ shows one broad peak, which is composed of many unresolved peaks. Account for the spectrum of $(CH_3)_3N:BCl_3$. Why is the spectrum of the BF_3 derivative so different from that of the BCl_3 adduct? (See Appendix 7 for natural abundances and nuclear spins of the isotopes of the elements.)

INDEPENDENT STUDIES

A. Measure the gas phase infrared spectra of BF_3 and $(CH_3)_3N$ and make band assignments. (See Experiment 10.)
B. Measure the mass spectrum of $(CH_3)_3N:BF_3$ and make assignments to all ion fragments. (G. F. Lanthier and J. M. Miller, *J. Chem. Soc. (A)*, 346 (1971).)
C. Do a vacuum line tensimetric titration of $(CH_3)_3N$ with BF_3 to establish the existence of both $(CH_3)_3N:BF_3$ and $(CH_3)_3N:2BF_3$ at low temperature. (J. L. Mills and L. C. Flukinger, *J. Chem. Educ.*, *50*, 636 (1973).)
D. Prepare and characterize $(CH_3)_3N:BCl_3$ or $(CH_3)_3N:BBr_3$.

E. Prepare and characterize an amine adduct of the Lewis acid SiF$_4$.
(B. J. Aylett, I. A. Ellis, and C. J. Porritt, *J. Chem. Soc. (Dalton)*, 1953
(1972).)

REFERENCES

(CH$_3$)$_3$N:BF

R. L. Amster and R. C. Taylor, *Spectrochim. Acta, 20,* 1487 (1964). Infrared spectra of
(CH$_3$)$_3$N:BF$_3$ and other adducts.

P. S. Bryan and R. L. Kuczkowski, *Inorg. Chem., 10,* 200 (1971). Microwave spectrum of
gaseous (CH$_3$)$_3$N:BF$_3$ and molecular structure.

A. B. Burg and A. A. Green, *J. Am. Chem. Soc., 65,* 1838 (1943). Preparation and properties of (CH$_3$)$_3$N:BF$_3$.

J. S. Hartman and J. M. Miller, *Inorg. Chem., 13,* 1467 (1974). ^{11}B and ^{19}F nmr spectra
of (CH$_3$)$_3$N:BF$_3$ and related adducts.

A. R. Katritzky, *J. Chem. Soc.,* 2049 (1959). Infrared spectra of (CH$_3$)$_3$N:BF$_3$ and other
adducts.

Vacuum Techniques

R. E. Dodd and P. L. Robinson, *Experimental Inorganic Chemistry,* Elsevier, New York,
1954, Chapter 2. Fundamentals of vacuum line construction and use.

S. Dushman, *Scientific Foundations of Vacuum Technique,* John Wiley and Sons, New York,
1949. Theory and practice of vacuum line use.

S. Jnanananda, *High Vacua,* D. VanNostrand Co., New York, 1947. Principles, production
and measurement of high vacua.

R. T. Sanderson, *Vacuum Manipulation of Volatile Compounds,* John Wiley and Sons, New
York, 1948. Design and construction of practical high vacuum apparatus.

D. F. Shriver, *The Manipulation of Air-Sensitive Compounds,* McGraw-Hill, New York, 1969.
Excellent coverage of the construction and use of vacuum lines.

J. Yarwood, *High Vacuum Technique,* 4th Ed., Chapman and Hall Ltd., London, 1967.
Vacuum line components and design.

Lewis Acid-Base Reactions

H. S. Booth and D. R. Martin, *Boron Trifluoride and Its Derivatives,* John Wiley and Sons,
New York, 1949. All aspects of the physical and chemical properties of BF$_3$.

D. R. Martin and J. M. Canon, in *Friedel-Crafts and Related Reactions, Vol. 1,* G. A. Olah,
Ed., Interscience Publishers, New York, 1963, p. 399. A comprehensive summary
of acid-base adducts of the boron halides, BX$_3$.

A. G. Massey in *Advances in Inorganic Chemistry and Radiochemistry, Vol. 10,* H. J. Emeleus
and A. G. Sharpe, eds., Academic Press, New York, 1967, p. 1. Preparations, structures, and chemistry of the boron halides.

F. G. A. Stone, *Chem. Revs., 58,* 101 (1958). Stabilities of numerous acid-base adducts of
the group III elements.

$(CH_3)_3CNH_2:BH_3$

The syntheses and reactions of the group of compounds known as the boron hydrides have been and continue to be active areas of research. The structures of the compounds B_2H_6, B_4H_{10}, B_5H_9, B_5H_{11}, B_9H_{15}, $B_{10}H_{14}$, $B_{10}H_{10}^{2-}$, $B_{12}H_{12}^{2-}$, and many others have only recently been established. Since this experiment is concerned only with the simplest of the boranes, B_2H_6, the almost incredible variety of structures and reactions of the higher boranes is left to the reader to pursue through the references given at the end of the chapter.

Diborane, B_2H_6, is most conveniently prepared from the reaction of sodium borohydride, $NaBH_4$ (frequently called sodium hydroborate), with BF_3 in ether.

$$3NaBH_4 + 4BF_3 \rightarrow 3NaBF_4 + 2B_2H_6$$

At room temperature it is a gas with the molecular structure,

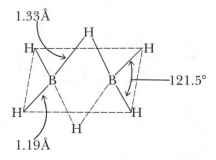

The two bridging H atoms lie above and below the plane described by the four terminal H atoms and the two B atoms. While the terminal H atoms presumably bond to the B atoms with normal two-electron bonds, each of the bridging H atoms must bond equally to two B atoms. Since the H atom has only one low energy atomic orbital, the 1s orbital, its bonding to both B atoms may be accounted for in terms of a three-center molecular orbital.

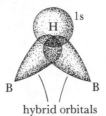

hybrid orbitals

Such molecular orbitals, consisting of the simultaneous overlap of three different atomic orbitals, are common to most of the boron hydrides. Although B$_2$H$_6$ exists largely in the bridged form, it does dissociate to a very small extent to give BH$_3$.

$$B_2H_6 \rightleftharpoons 2BH_3$$

Borane, BH$_3$, is the analog of other Lewis acids such as BF$_3$ (see Experiment 19) and forms numerous acid-base adducts with donor molecules such as amines and phosphines. Primary, secondary, and tertiary amines react with diborane to yield 1:1 adducts; characteristic is the reaction of trimethylamine:

$$2(CH_3)_3N + B_2H_6 \rightarrow 2(CH_3)_3N{:}BH_3$$

The geometry around both the N and B atoms is approximately tetrahedral, and the adduct can be considered to be very similar to that of the all-carbon analog (CH$_3$)$_3$CCH$_3$.

Historically, borane adducts have been prepared from B$_2$H$_6$, which inflames in air and is immediately hydrolyzed by water. In addition to these hazards B$_2$H$_6$ is exceedingly toxic. For these reasons, a more convenient and less dangerous route to the borane adducts was sought. It was found in the reaction of NaBH$_4$ with the desired alkylammonium salt:

$$R_3NH^+Cl^- + NaBH_4 \rightarrow R_3N{:}BH_3 + NaCl + H_2$$

These reactions proceed under mild conditions, giving good yields of the borane adducts. The particular reaction to be carried out in this experiment is that of t-butylammonium chloride; this salt may be simply prepared by bubbling gaseous HCl into an ether solution of t-butylamine:

$$(CH_3)_3CNH_2 + HCl \xrightarrow{\;(C_2H_5)_2O\;} (CH_3)_3CNH_3^+Cl^-$$

The borane adduct is formed according to the following equation:

$$(CH_3)_3CNH_3^+Cl^- + NaBH_4 \rightarrow (CH_3)_3CNH_2{:}BH_3 + NaCl + H_2$$

The product, $(CH_3)_3CNH_2{:}BH_3$, is a white solid (m.p., 96°C) that is stable toward air and water at room temperature. It will be characterized by its infrared spectrum. The proton nmr spectrum (see Experiment 14, p. 137) of $(CH_3)_3CNH_2{:}BH_3$ consists of one sharp peak corresponding to the $-CH_3$ protons at approximately -1.2 ppm relative to the internal standard, tetramethylsilane. The absorptions of the protons on the N and B atoms are broadened so greatly by the nuclear quadrupole moments of ^{14}N and ^{11}B that they are not visible in the spectrum.

Borane adducts undergo many reactions. Some involve the replacement of H by Cl in the BH_3 portion of the molecule. This may be accomplished by reaction with gaseous HCl.

$$(CH_3)_3CNH_2{:}BH_3 + HCl \xrightarrow{\quad (C_2H_5)_2O \quad} (CH_3)_3CNH_2{:}BH_2Cl + H_2$$

This reaction again illustrates the hydridic, H^-, nature of the H atoms attached to boron. They readily combine with protonic, H^+, hydrogen atoms to produce H_2. The analogous reaction with HF replaces all three hydrogens on the boron.

$$(CH_3)_3CNH_2{:}BH_3 + 3HF \rightarrow (CH_3)_3CNH_2{:}BF_3 + 3H_2$$

The same adduct may be prepared by reacting $(CH_3)_3CNH_2$ and BF_3 directly.

Trialkylamine-boranes react with olefins to form the product in which the B and H have added across the olefinic bond.

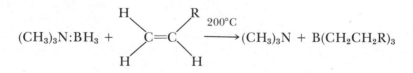

$$(CH_3)_3N{:}BH_3 + \quad \text{(olefin)} \quad \xrightarrow{200°C} (CH_3)_3N + B(CH_2CH_2R)_3$$

The trialkylborane, $B(CH_2CH_2R)_3$, is a sufficiently weak Lewis acid that it may be liberated from the trimethylamine without difficulty. The ability of a borane to add across an olefinic bond is the basis for the widespread use of $NaBH_4$ in organic chemistry.

Amine-boranes decompose with the loss of H_2 at high temperatures to a variety of products, depending upon the particular reactant and the conditions. Heating $(CH_3)_2NH{:}BH_3$, for example, gives the aminoborane $(CH_3)_2NBH_2$, having a planar configuration around the B atom.

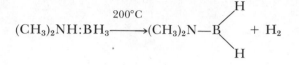

The analogous monomethylamine-borane, when heated at 100°C, yields the B–N analog of 1,3,5–trimethylcyclohexane. Further loss of H_2 occurs at roughly 200°C to produce the unsaturated cyclic ring compound

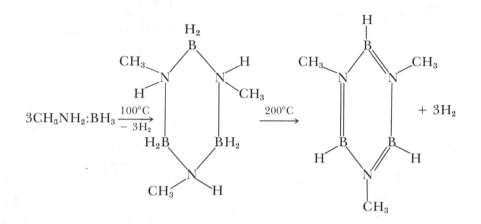

called 1,3,5–trimethylborazine. Its carbon analog is mesitylene, 1,3,5–$(CH_3)_3C_6H_3$. The parent borazine, $B_3N_3H_6$,

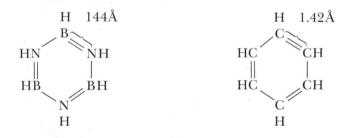

has been studied in some detail and structurally resembles the isoelectronic benzene. For this reason borazine is sometimes called "inorganic benzene." Like benzene, it has a planar hexagonal structure with relatively short B–N bonds indicative of B–N π-bonding. Despite the physical similarities, their chemical reactivities are quite different. The π-system of benzene is relatively inert to reaction, whereas borazine adds hydrogen halides, such as HCl, to give the saturated cyclic aminoborane.

$$B_3N_3H_6 + 3HCl \rightarrow$$

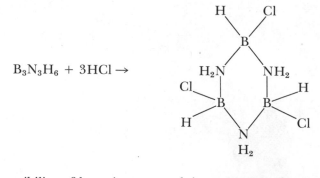

The susceptibility of borazine to attack is probably related to the polar nature of the B–N bond.

EXPERIMENTAL SECTION

In contrast to many hydride compounds, such as NaH, CaH_2, or $LiAlH_4$, which react explosively with water, $NaBH_4$ is stable in neutral or alkaline aqueous solutions, but hydrolyzes rapidly in acidic solution.

$(CH_3)_3CNH_3{}^+Cl^-$

Dissolve 2.5 ml (1.7 g, 23 mmoles) of $(CH_3)_3CNH_2$ in approximately 20 ml of anhydrous ethyl ether. Bubble gaseous HCl from a lecture bottle compressed gas cylinder into the solution until precipitation of $(CH_3)_3CNH_3{}^+Cl^-$ is complete. Suction-filter (Figure 1–1) the product on a medium frit, wash with a few milliliters of ether, and suck dry on the frit. Determine the yield. Although many alkylammonium chloride salts are very hygroscopic, $(CH_3)_3CNH_3{}^+Cl^-$ is not; it need not be stored in a desiccator except when the humidity is high.

$(CH_3)_3CNH_2{:}BH_3$

Assemble the apparatus shown in Figure 20–1 and lubricate the stirring shaft bearing with glycerin. The drying tube is necessary only if the atmospheric relative humidity is very high. Add 1.3 g (11.8 mmoles) of $(CH_3)_3CNH_3{}^+Cl^-$ and 15 ml of tetrahydrofuran to the 250 ml 3-neck flask. (The tetrahydrofuran may be used as obtained commercially unless it contains large amounts of water. Then it should be dried over NaOH and distilled.) To the stirred suspension, add 0.20 g (5.3 mmoles) of powdered $NaBH_4$, whereupon H_2 gas will be evolved. Add an additional 10 or 15 ml of tetrahydrofuran and continue stirring the solution

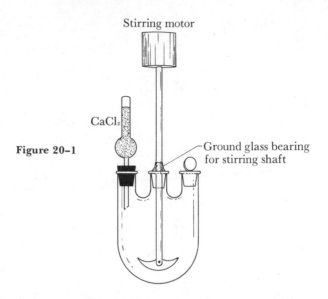

Figure 20–1

Stirring motor

CaCl$_2$

Ground glass bearing for stirring shaft

for about 2 hours at room temperature. Suction-filter (Figure 1–1) off the NaCl and unreacted excess (CH$_3$)$_3$CNH$_3$$^+Cl^-$. After the filtration, disconnect the rubber vacuum tubing from the filter flask before the water flow is turned off. This will prevent water from backing up into the trap (Figure 1–1). Then reattach the vacuum tubing to the flask and place a rubber stopper in the top of the flask. Turn on the water aspirator and evaporate the tetrahydrofuran solution to dryness, leaving the product, (CH$_3$)$_3$CNH$_2$:BH$_3$. The evaporation proceeds more rapidly if the flask is swirled in a beaker of warm water. Before turning off the water aspirator, first disconnect the vacuum tubing at the flask. The (CH$_3$)$_3$CNH$_2$:BH$_3$ obtained is usually of high purity and need not be recrystallized. Determine its melting point to confirm this. If the melting range is greater than 4°C, it should be recrystallized by dissolving it in a minimum amount (1 to 2 ml) of benzene and adding hexane (~20 ml) until precipitation is complete. Suction-filter and dry the (CH$_3$)$_3$CNH$_2$:BH$_3$. Redetermine its melting point. Calculate the yield. Measure the infrared spectrum of the product in CCl$_4$ solution (see Experiment 14, p. 136).

REPORT

Include the following:
1. Percentage yields of (CH$_3$)$_3$CNH$_3$$^+Cl^-$ and (CH$_3$)$_3$CNH$_2$:BH$_3$.
2. Melting point of (CH$_3$)$_3$CNH$_2$:BH$_3$.
3. Infrared spectrum of (CH$_3$)$_3$CNH$_2$:BH$_3$ with assignments to vibrational modes in the molecule. Compare the B—H, C—H, and N—H stretching frequencies and account for their differences.

QUESTIONS

1. Propose a mechanism for the reaction of $NaBH_4$ with $(CH_3)_3CNH_3^+Cl^-$.
2. Suggest a method of establishing the presence of boron in your product, $(CH_3)_3CNH_2:BH_3$.
3. Earlier it was noted that B_2H_6 inflames in air and rapidly hydrolyzes in water. Write balanced equations for these reactions.
4. Write a balanced equation for the hydrolysis of $NaBH_4$ in acidic solution.
5. Draw structures of the following: $NaBF_4$, $NaBH_4$, $B(CH_3)_3$, and BF_3.
6. Tetrahydrofuran (THF) forms an adduct with BH_3, $THF:BH_3$. Draw the structure of this compound. Is there any evidence from this experiment that would suggest that THF coordinates more or less strongly to BH_3 than does $(CH_3)_3CNH_2$?
7. Account for the fact that $LiBH_4$ is more soluble in THF than is $NaBH_4$. Would you expect $LiBH_4$ or $NaBH_4$ to give better yields in the present experiment? Why?
8. If you wished to carry out a reaction of the type,

$$(CH_3)_3CNH_2:BH_3 + \text{amine} \rightarrow \text{amine}:BH_3 + (CH_3)_3CNH_2,$$

what amine would you choose and what reaction conditions would you use to drive the reaction to completion?

INDEPENDENT STUDIES

A. Prepare and characterize other amine-borane adducts, such as $(CH_3)_2NH:BH_3$ and $(CH_3)_3N:BH_3$. (H. Nöth and H. Beyer, *Chem. Ber.*, *93*, 928 (1960). K. C. Nainan and G. E. Ryschkewitsch, *Inorganic Syntheses*, *15*, 122 (1974).)
B. Using your $(CH_3)_3CNH_2:BH_3$, prepare and characterize $(CH_3)_3CNH_2:BH_2X$ (where X = F, Cl, Br, or I). (H. Nöth and H. Beyer, *Chem. Ber.*, *93*, 2251 (1960). G. E. Ryschkewitsch and J. W. Wiggins, *Inorganic Syntheses*, *12*, 116 (1970).)
C. Prepare and characterize pyridine:BH_3 and $[(\text{pyridine})_2BH_2^+]I^-$. (G. E. Ryschkewitsch, *J. Amer. Chem. Soc.*, *89*, 3145 (1967). K. C. Nainan and G. E. Ryschkewitsch, *Inorg. Chem.*, *7*, 1316 (1968).)
D. Determine the mass spectrum of $(CH_3)_3CNH_2:BH_3$ and make assignments to all ion fragments.
E. Prepare the interesting tridentate ligand, hydrotris(1-pyrazolyl) borate, $HB(C_3H_3N_2)_3^-$. (S. Trofimenko, *Inorganic Syntheses*, *12*, 102 (1970).)

REFERENCES

RNH$_2$:BH$_3$

H. Nöth and H. Beyer, *Chem. Ber., 93,* 928 (1960). Preparation and characterization of borane adducts.

B. Rice, R. J. Galiano, and W. J. Lehmann, *J. Phys. Chem., 61,* 1222 (1957). Infrared study of (CH$_3$)$_3$N:BH$_3$.

R. C. Taylor in *Boron-Nitrogen Chemistry,* Advances in Chemistry Series, No. 42, American Chemical Society, Washington D. C., 1964, p. 59. Infrared spectra of borane adducts.

Boron Hydride Chemistry

R. M. Adams, ed., *Boron, Metallo-boron Compounds and Boranes,* Interscience Publishers, New York, 1964. Chapters on borax, the borates, hydroborates, boron hydrides, and toxicology of B compounds.

M. P. Brown, R. H. Heseltine, and L. H. Sutcliffe. *J. Chem. Soc. (A),* 612 (1968). Products resulting from heating RNH$_2$:BH$_3$ compounds.

R. A. Geanangel and S. G. Shore, *Preparative Inorganic Reactions,* W. L. Jolly, ed., Interscience Publishers, New York, 1966, p. 123. An excellent review of boron-nitrogen compounds.

W. N. Lipscomb, *Boron Hydrides,* W. A. Benjamin, Inc., New York, 1963, paperback. Introduction to structures, bonding, and reactions of the boron hydrides.

E. L. Muetterties and W. H. Knoth, *Polyhedral Boranes,* Marcel Dekker, Inc., New York, 1968. Recent developments in structure, bonding and reactions of the higher boranes.

K. Niedenzu and J. W. Dawson, *Boron-Nitrogen Compounds,* Springer-Verlag, Berlin-New York, 1965. A very useful review of all types of B–N compounds.

NMR and Mass Spectra

This section includes nmr and mass spectra of compounds synthesized in the experiments. They are given here for the student to interpret as part of his characterization of the compounds. Discussions of nuclear magnetic resonance and mass spectrometry are given in Experiment 14.

Proton NMR Spectra

These spectra are recordings taken directly from the Varian A60 nmr spectrometer. The only change is that the integrals of the peak areas, which are normally recorded in blue ink, have been changed to dashed (----) lines. Tetramethylsilane (TMS) was added to all samples and is shown at $\delta = 0$ ppm in all spectra.

Mass Spectra

These spectra were obtained directly from the recorder of an Atlas CH4 mass spectrometer. All spectra were recorded using 70 eV electrons. This is a very high energy, sufficient to produce a great deal of fragmentation, as noted from the spectra. To obtain reasonable peak heights, the recorder voltage is changed in different mass ranges. A recorder voltage change is indicated by Γ in the spectra; this indicates that peaks recorded to the right of the arrow are recorded at the indicated voltage. The voltage is a measure of the voltage required to give a full-scale deflection. To compare two peaks of the same height, one of which is recorded at 0.3 v. and the other at 0.1 v., the height of the peak recorded at 0.3 v. should be multiplied by 3 and then compared with that of the peak measured at 0.1 v. The mass scale in a.m.u. is self-explanatory.

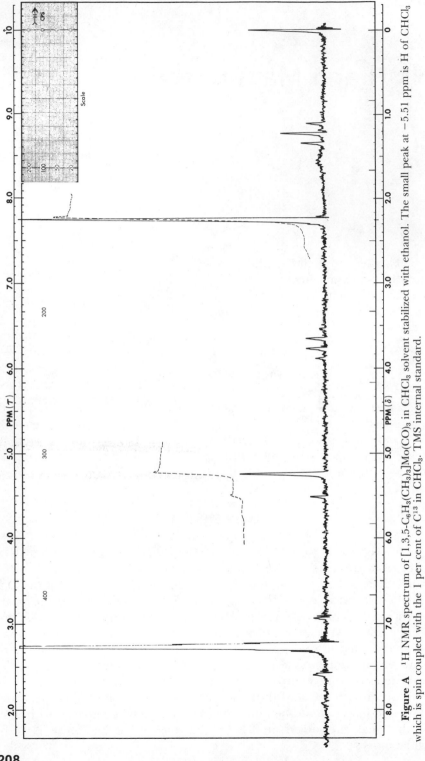

Figure A ^{1}H NMR spectrum of $[1,3,5\text{-}C_6H_3(CH_3)_3]Mo(CO)_3$ in $CHCl_3$ solvent stabilized with ethanol. The small peak at -5.51 ppm is H of $CHCl_3$ which is spin coupled with the 1 per cent of C^{13} in $CHCl_3$. TMS internal standard.

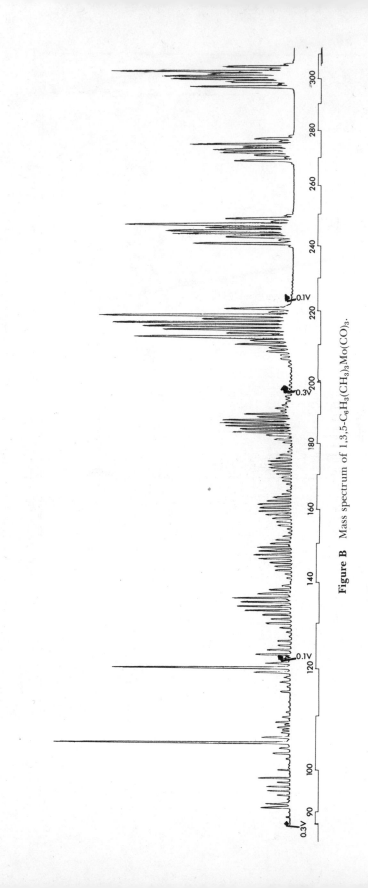

Figure B Mass spectrum of 1,3,5-$C_6H_3(CH_3)_3Mo(CO)_3$.

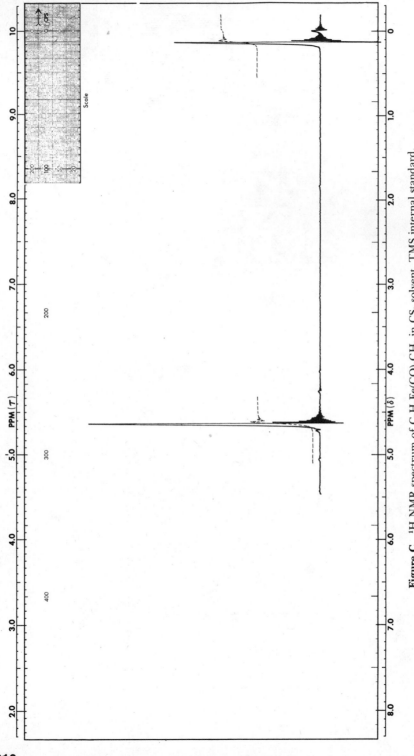

Figure C ^1H NMR spectrum of $C_5H_5Fe(CO)_2CH_3$ in CS_2 solvent. TMS internal standard.

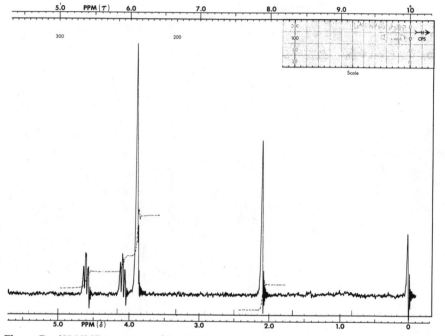

Figure D   ¹H NMR spectrum of $Fe(C_5H_5)(C_5H_4COCH_3)$ in C_6H_6 solvent with TMS internal standard. The C_6H_6 absorption occurs at -7.2 ppm and is not shown.

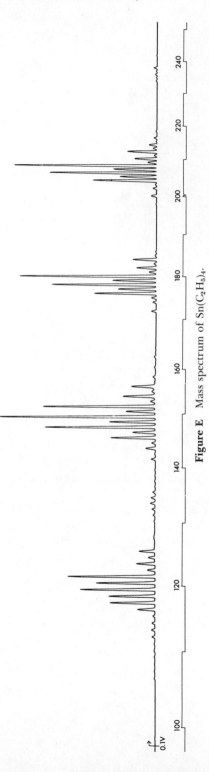

Figure E Mass spectrum of $Sn(C_2H_5)_4$.

Appendices

APPENDIX 1 **APPENDIX 1** **Approximate Concentrations of Commercial, Reagent-grade Acids and Bases**

Acid or Base	% by Weight	Molarity
Acetic acid (glacial)	99.8	17.4
Ammonia (aqueous)	29	14.8
Hydrochloric acid	37	12.0
Nitric acid	70	15.9
Perchloric acid	70	11.7
Phosphoric acid	85	14.7
Sulfuric acid	96	18

APPENDIX 2 **Expected Molar Conductance (Λ_M) Ranges[a] for 2, 3, 4, and 5 Ion Electrolytes ($\sim 10^{-3}$ M) in Some Common Solvents at 25°C[b]**

Solvent	Dielectric Constant	Electrolyte Types			
		1:1	2:1	3:1	4:1
Water[c]	78.4	118–131	235–273	408–435	~560
Nitromethane	35.9	75–95	150–180	220–260	290–330
Nitrobenzene	34.8	20–30	50–60	70–82	90–100
Acetone	20.7	100–140	160–200		
Acetonitrile	36.2	120–160	220–300	340–420	
Dimethylformamide	36.7	65–90	130–170	200–240	
Methanol	32.6	80–115	160–220		
Ethanol	24.3	35–45	70–90		

[a]Units on all molar conductances are ohm^{-1}cm^{2}mole^{-1}.
[b]W. J. Geary, *Coord. Chem. Rev.*, 7, 81 (1971).
[c]M. Sneed and J. Maynard, *General Inorganic Chemistry*, Van Nostrand, New York, 1942, p. 813.

213

APPENDIX 3 Transmission Ranges of Cell Materials for Infrared and Ultraviolet-Visible Spectroscopy

Infrared Region[a]

Material	Useful Transmission Range	Comments
NaCl	5000–625 cm^{-1}	low cost, fogs slowly, but is water soluble
KBr	5000–400 cm^{-1}	properties similar to NaCl
CaF$_2$	5000–1110 cm^{-1}	insoluble in water, resists many acids and bases, difficult to polish
BaF$_2$	5000–830 cm^{-1}	properties similar to CaF$_2$
CsBr	5000–250 cm^{-1}	water soluble and very hygroscopic, difficult to polish
CsI	5000–165 cm^{-1}	properties similar to CsBr, but less hygroscopic
AgCl	5000–435 cm^{-1}	insoluble in water, darkens in light, difficult to polish
AgBr	5000–285 cm^{-1}	properties similar to AgCl
KRS-5 (TlBr, TlI)	5000–250 cm^{-1}	slightly soluble in water, toxic
Irtran-2 (ZnS)	5000–835 cm^{-1}	insoluble in water, rugged
Sapphire (Al$_2$O$_3$)	5000–780 cm^{-1}	insoluble in water, hard
Quartz (SiO$_2$)	5000–2700 cm^{-1}	insoluble in water, rugged
Ge	5000–600 cm^{-1}	insoluble in water, brittle, high reflection losses
Si	5000–660 cm^{-1}	properties similar to Ge
Polyethylene	625–33 cm^{-1}	insoluble in water, low cost, difficult to clean

Ultraviolet-Visible Region

Pyrex	320–2500 nm
Vycor	320–2500 nm
Standard Silica	220–2600 nm
NIR Silica	220–3500 nm
Far-UV Silica	160–2600 nm

[a]R. G. J. Miller, Ed., *Laboratory Methods in Infrared Spectroscopy*, 2nd Ed., Heyden and Son, Ltd., London, 1972.

APPENDIX 4 Properties of Commercial Glasses

	Soda-Lime (Soft; Flint; TEKK; EXAX)	Borosilicate (Kimax; Pyrex)	96% Silica (Vycor)	Fused Silica
Strain point[a] (°C)	480	510	820	1020
Annealing point[b] (°C)	520	550	910	1120
Softening point[c] (°C)	700	820	1500	1710
Working point[d] (°C)	1010	1245		
Linear coefficient of expansion[e]	9.3	3.3	0.8	0.6
Refractive index (n$_D$)	1.52	1.474	1.458	1.54
Density (g/cm^3)	2.5	2.23	2.18	2.65

[a]Point at which strain is relieved after 4 hr.
[b]Point at which strain is relieved in 15 min.
[c]Point at which sagging begins.
[d]Point at which glass can be easily manipulated and blown by mouth.
[e]Parts per million per °C between 0 and 300°C.
(A. J. Gordon and R. A. Ford, *The Chemist's Companion,* John Wiley & Sons, New York, 1972.)

APPENDIX 5 Common Solvents and Their Properties

Name, Formula	b.p.,[a] °C	m.p.,[a] °C	ϵ[b]	
Acetic acid, CH_3CO_2H	118	16.6	6.2	
Acetone, $(CH_3)_2CO$	56.2	−95.4	20.7	
Acetonitrile, CH_3CN	81.6	−45.7	36.2	
Ammonia, NH_3	−33.4	−77.7	27	(−60)
Benzene, C_6H_6	80.1	5.5	2.3	(20)
Carbon disulfide, CS_2	46.2	−111.5	2.6	
Carbon tetrachloride, CCl_4	76.5	−23	2.2	
Chlorobenzene, C_6H_5Cl	132	−46	5.6	
Chloroform, $CHCl_3$	61.7	−63.5	4.7	
Cyclohexane, $(CH_2)_6$	80.7	6.5	2.0	
cis-Decalin	196	−43	2.2	(20)
trans-Decalin	187	−30	2.2	(20)
1,2-Dichloroethane, ClH_2CCH_2Cl	83.5	−35	10.4	
o-Dichlorobenzene, $C_6H_4Cl_2$	180	−17	9.9	
Dichloromethane, CH_2Cl_2	40	−95	8.9	
Diethyl ether, $(C_2H_5)_2O$	34.5	−116	4.3	(20)
Diglyme, $(CH_3OCH_2CH_2)_2O$	160			
N,N-Dimethylformamide, $HCON(CH_3)_2$	152	−61	36.7	
Dimethyl sulfoxide, $(CH_3)_2SO$	189	18.4	49	
Dinitrogen tetroxide, N_2O_4	21.3	−12.3	2.4	(18)
1,4-Dioxane, $O(CH_2CH_2)_2O$	102	11.8	2.2	
Ethanol, CH_3CH_2OH	78.3	−114	24.3	
Ethyl acetate, $CH_3CO_2C_2H_5$	77.1	−83.6	6.0	
Formamide, $HCONH_2$	193	2.6	110	
Glyme, $CH_3OCH_2CH_2OCH_3$	83	−58		
Hexamethylphosphoramide, $[(CH_3)_2N]_3PO$	233	7.2	30	(20)
n-Hexane, $CH_3(CH_2)_4CH_3$	69	−95	1.9	(20)
Hydrogen cyanide, HCN	26	−14	115	(20)
Hydrogen fluoride, HF	19.5	−89.4	84	(0)
Methanol, CH_3OH	64.5	−97.5	32.6	
Nitric acid, HNO_3	82.6	−41.6		
Nitrobenzene, $C_6H_5NO_2$	211	5.8	35	(30)
Nitromethane, CH_3NO_2	101	−28.5	38.6	
n-Pentane, $CH_3(CH_2)_3CH_3$	36.1	−130	1.8	
Phosphoric acid, H_3PO_4	213	42.4		
Pyridine, C_5H_5N	116	−42	12.3	
Sulfur dioxide, SO_2	−10.1	−75.5	15.4	(0)
Sulfuric acid, H_2SO_4	~305	10.4	101	
1,1,2,2-Tetrachloroethane, $Cl_2HCCHCl_2$	146	−36	8.2	(20)
Tetrachloroethylene, $Cl_2C{=}CCl_2$	121	−19	2.5	
Tetrahydrofuran, $(CH_2)_4O$	66	−65	7.3	
Thionyl chloride, $SOCl_2$	80	−105	9.2	(20)
Toluene, $C_6H_5CH_3$	111	−95	2.4	
Water, H_2O	100.0	0	78.5	

[a]Boiling points and melting points are at 760 torr (mm Hg) unless indicated otherwise in parentheses.

[b]Dielectric constants are at 25°C unless indicated otherwise in parentheses.

APPENDIX 6   Drying (Dehydrating) Agents[a]

Agent	Properties
Al_2O_3 or SiO_2	Very high capacity and fast; good for solvents or gases; may be regenerated at high temperature under vacuum. Equilibrium water vapor pressure above them is $\sim 10^{-3}$ mm Hg.
BaO or CaO	Forms $Ba(OH)_2$ or $Ca(OH)_2$; cannot be used with solvents sensitive to base. Slow but efficient. Water vapor pressures above BaO and CaO are 7×10^{-4} and 3×10^{-3} mm Hg, respectively.
$CaCl_2$	Very fast, but not efficient (0.2 mm Hg water vapor pressure).
CaH_2	Forms H_2 and $Ca(OH)_2$. Excellent efficiency ($<10^{-5}$ mm Hg water vapor pressure). Do not use with halogenated solvents or those with active groups such as aldehydes. Slower but just as efficient as $LiAlH_4$.
$CaSO_4$	Low capacity but fast. Water vapor pressure above it is 5×10^{-3} mm Hg. Can be regenerated. Sold commercially as "Drierite."
H_2SO_4 (concentrated)	Fast and has high capacity. Water vapor pressure is $\sim 3 \times 10^{-3}$ mm Hg. Cannot be used with basic compounds.
KOH	Fast and very high capacity. Water vapor pressure of $\sim 2 \times 10^{-3}$ mm Hg. Especially good for amines.
$LiAlH_4$	Forms H_2, LiOH, and $Al(OH)_3$. Use only with inert solvents, since it reacts with a variety of organic functional groups. Very efficient.
$MgSO_4$	Fast and fair capacity; an excellent general drying agent.
Molecular sieves[b] (Types 3A and 4A)	Fast and high capacity. Very efficient ($\sim 1 \times 10^{-3}$ mm Hg water vapor pressure). May be regenerated at 350°C in vacuum.
Na	Forms H_2 and NaOH. Use only with inert solvents; explosive with halogenated organic solvents.
Na_2SO_4	Slow and inefficient, but high capacity. Good for pre-drying; may be regenerated at 150°C.
P_4O_{10}	Forms phosphoric acids. Very fast and efficient (water vapor pressure of 2×10^{-5} mm Hg).

[a]A. J. Gordon and R. A. Ford, *The Chemist's Companion,* John Wiley & Sons, New York, 1972, p. 445. D. F. Shriver, *The Manipulation of Air-Sensitive Materials,* McGraw-Hill, New York, 1969, p. 194.
[b]D. W. Breck, *J. Chem. Educ., 41,* 678 (1964).

APPENDIX 7   Natural Abundances and Nuclear Spins of Stable Isotopes

Isotope	Natural Abundance,[a] %	Nuclear Spin,[c,d] I	Isotope	Natural Abundance,[a] %	Nuclear Spin,[c,d] I
^{1}H	99.985	1/2	^{50}V	0.25	6
^{2}H	0.015	1	^{51}V	99.75	7/2
^{4}He	100		^{50}Cr	4.31	
^{6}Li	7.42	1	^{52}Cr	83.76	
^{7}Li	92.58	3/2	^{53}Cr	9.55	3/2
^{9}Be	100	3/2	^{54}Cr	2.38	
^{10}B	19.7	3	^{55}Mn	100	5/2
^{11}B	80.3	3/2	^{54}Fe	5.84	
^{12}C	98.892		^{56}Fe	91.68	
^{13}C	1.108	1/2	^{57}Fe	2.17	
^{14}N	99.635	1	^{58}Fe	0.31	
^{15}N	0.365	1/2	^{59}Co	100	7/2
^{16}O	99.759		^{58}Ni	67.76	
^{17}O	0.037	5/2	^{60}Ni	26.16	
^{18}O	0.204		^{61}Ni	1.25	
^{19}F	100	1/2	^{62}Ni	3.66	
^{20}Ne	90.92		^{64}Ni	1.16	
^{21}Ne	0.257		^{63}Cu	69.1	3/2
^{22}Ne	8.82		^{65}Cu	30.9	3/2
^{23}Na	100	3/2	^{64}Zn	48.89	
^{24}Mg	78.60		^{66}Zn	27.81	
^{25}Mg	10.11	5/2	^{67}Zn	4.11	5/2
^{26}Mg	11.29		^{68}Zn	18.56	
^{27}Al	100	5/2	^{70}Zn	0.62	
^{28}Si	92.18		^{69}Ga	60.2	3/2
^{29}Si	4.71	1/2	^{71}Ga	39.8	3/2
^{30}Si	3.12		^{70}Ge	20.55	
^{31}P	100	1/2	^{72}Ge	27.37	
^{32}S	95.0		^{73}Ge	7.67	9/2
^{33}S	0.76	3/2	^{74}Ge	36.74	
^{34}S	4.22		^{76}Ge	7.67	
^{35}Cl	75.53	3/2	^{75}As	100	3/2
^{37}Cl	24.47	3/2	^{74}Se	0.87	
^{40}Ar	99.60[b]		^{76}Se	9.02	
^{39}K	93.22	3/2	^{77}Se	7.58	1/2
^{41}K	6.77	3/2	^{78}Se	23.52	
^{40}Ca	96.97		^{80}Se	49.82	
^{42}Ca	0.64		^{82}Se	9.19	
^{44}Ca	2.06		^{79}Br	50.52	3/2
^{45}Sc	100	7/2	^{81}Br	49.48	3/2
^{46}Ti	.7.99		^{84}Kr	56.90[b]	
^{47}Ti	7.32	5/2	^{85}Rb	72.15	5/2
^{48}Ti	73.99		^{87}Rb	27.85	3/2
^{49}Ti	5.46	7/2	^{84}Sr	0.56	
^{50}Ti	5.25		^{86}Sr	9.86	

[a]*Lange's Handbook of Chemistry,* J. A. Dean, ed., 11th Ed., McGraw-Hill, New York, 1973.

[b]Other isotopes not listed here.

[c]In multiples of $h/2\pi$.

[d]R. Chang, *Basic Principles of Spectroscopy,* McGraw-Hill, New York, 1971, p. 286.

Table continued on following page.

Appendices

APPENDIX 7　Natural Abundances and Nuclear Spins of Stable Isotopes *(Continued)*

Isotope	Natural Abundance,[a] %	Nuclear Spin,[c,d] I	Isotope	Natural Abundance,[a] %	Nuclear Spin,[c,d] I
^{87}Sr	7.02	9/2	^{110}Cd	12.39	
^{88}Sr	82.56		^{111}Cd	12.75	1/2
^{89}Y	100	1/2	^{112}Cd	24.07	
^{90}Zr	51.46		^{113}Cd	12.26	1/2
^{91}Zr	11.23	5/2	^{114}Cd	28.86	
^{92}Zr	17.11		^{116}Cd	7.58	
^{94}Zr	17.40		^{113}In	4.23	9/2
^{96}Zr	2.80		^{115}In	95.77	
^{93}Nb	100	9/2	^{112}Sn	0.95	
^{92}Mo	15.86		^{114}Sn	0.65	
^{94}Mo	9.12		^{115}Sn	0.34	1/2
^{95}Mo	15.70	5/2	^{116}Sn	14.24	
^{96}Mo	16.50		^{117}Sn	7.57	1/2
^{97}Mo	9.45	5/2	^{118}Sn	24.01	
^{98}Mo	23.75		^{119}Sn	8.58	1/2
^{100}Mo	9.62		^{120}Sn	32.97	
^{96}Ru	5.46		^{122}Sn	4.71	
^{98}Ru	1.87		^{124}Sn	5.98	
^{99}Ru	12.63	5/2	^{121}Sb	57.25	5/2
^{100}Ru	12.53		^{123}Sb	42.75	7/2
^{101}Ru	17.02	5/2	^{122}Te	2.46	
^{102}Ru	31.6		^{123}Te	0.87	1/2
^{104}Ru	18.87		^{124}Te	4.61	
^{103}Rh	100	1/2	^{125}Te	6.99	1/2
^{102}Pd	0.96		^{126}Te	18.71	
^{104}Pd	10.97		^{128}Te	31.79	
^{105}Pd	22.2	5/2	^{130}Te	34.49	
^{106}Pd	27.3		^{127}I	100	5/2
^{108}Pd	26.7		^{129}Xe	26.44[b]	1/2
^{110}Pd	11.8		^{132}Xe	26.89	
^{107}Ag	51.35	1/2	^{133}Cs	100	7/2
^{109}Ag	48.65	1/2	^{134}Ba	2.42	
^{106}Cd	1.22		^{135}Ba	6.59	3/2
^{108}Cd	0.88		^{136}Ba	7.81	

APPENDIX 7   Natural Abundances and Nuclear Spins of Stable Isotopes *(Continued)*

Isotope	Natural Abundance,[a] %	Nuclear Spin,[c,d] I	Isotope	Natural Abundance,[a] %	Nuclear Spin,[c,d] I
^{137}Ba	11.32	3/2	^{181}Ta	99.99	7/2
^{138}Ba	71.66		^{182}W	26.4	
^{139}La	99.91	7/2	^{183}W	14.4	1/2
^{138}Ce	0.25		^{184}W	30.6	
^{140}Ce	88.48		^{186}W	28.4	
^{142}Ce	11.07		^{185}Re	37.07	5/2
^{141}Pr	100	5/2	^{187}Re	62.93	5/2
^{142}Nd	27.13[b]		^{186}Os	1.59	
^{144}Nd	23.87		^{187}Os	1.64	
^{152}Sm	26.63[b]		^{188}Os	13.3	
^{154}Sm	22.53		^{189}Os	16.1	3/2
^{151}Eu	47.77	5/2	^{190}Os	26.4	
^{153}Eu	52.23	5/2	^{192}Os	41.0	
^{156}Gd	20.47[b]		^{191}Ir	38.5	3/2
^{158}Gd	24.9		^{193}Ir	61.5	3/2
^{160}Gd	21.9		^{192}Pt	0.78	
^{159}Tb	100	3/2	^{194}Pt	32.9	
^{162}Dy	25.53[b]		^{195}Pt	33.8	1/2
^{163}Dy	24.97	7/2	^{196}Pt	25.2	
^{164}Dy	28.18		^{198}Pt	7.19	
^{165}Ho	100	7/2	^{197}Au	100	3/2
^{166}Er	33.41[b]		^{198}Hg	10.02	
^{167}Er	22.94		^{199}Hg	16.84	1/2
^{168}Er	27.07		^{200}Hg	23.13	
^{169}Tm	100	1/2	^{201}Hg	13.22	3/2
^{172}Yb	21.82[b]		^{202}Hg	29.80	
^{174}Yb	31.84		^{204}Hg	6.85	
^{175}Lu	97.40	7/2	^{203}Tl	29.50	1/2
^{176}Lu	2.60		^{205}Tl	70.50	1/2
^{176}Hf	5.21		^{204}Pb	1.40	
^{177}Hf	18.56		^{206}Pb	25.1	
^{178}Hf	27.1		^{207}Pb	21.7	1/2
^{179}Hf	13.75		^{208}Pb	52.3	
^{180}Hf	35.22		^{209}Bi	100	9/2

APPENDIX 8   Common Adsorbents[a] for Column and Thin-Layer Chromatography

Silica, SiO_2

Also called silica gel. A general-purpose adsorbent useful for a broad range of organic and ionic compounds. Commercially available in the partially deactivated form (containing 10 to 20 per cent H_2O). Is a stronger adsorbent when activated by removing the water at 150° under vacuum.

Alumina, Al_2O_3

Commercially available in basic, neutral, and acidic forms. The basic form is useful in separating basic and neutral compounds that are stable to base. Neutral Al_2O_3 is less active but is effective in the separation of relatively strongly adsorbing groups such as ketones and esters. The acidic form is least active and is useful primarily for separating acids. All forms may be activated or deactivated by removing or adding H_2O.

Magnesium Silicate

A compound of MgO and SiO_2 sold under commercial names of Florisil, Magnesol, or Magnesium Trisilicate. A weaker adsorbent, similar to acidic Al_2O_3, useful in separating compounds with polar groups such as ketones and esters.

Magnesia, MgO

Similar to Al_2O_3 but better for separating olefinic and aromatic compounds. It has an unusual selectivity for unsaturated organics.

[a]A. J. Gordon and R. A. Ford, *The Chemist's Companion*, John Wiley & Sons, New York, 1972, p. 369. An excellent summary of conditions for different types of chromatography and many useful references.

APPENDIX 9   Baths for Cooling and Heating

I. Cold Baths
   A. Dry Ice (Solid CO_2)–Organic Solvent Baths[a, b]

Solvent	Bath Temperature (°C)
Carbon tetrachloride	−23°
Acetonitrile	−42°
Cyclohexanone	−46°
Chloroform	−61°
Acetone	−78°

[a]Small lumps of solid CO_2 are added to the solvent until a small excess of the solid is present. A. M. Phipps and D. N. Hume, *J. Chem. Educ.*, 45, 664 (1968).

[b]Baths useful in the −50° to 0°C temperature range have been prepared by adding solid CO_2 to various concentrations of $CaCl_2$–H_2O solutions. W. P. Bryan and R. H. Byrne, *J. Chem. Educ.*, 47, 361 (1970).

   B. Liquid Nitrogen–Organic Solvent Slush Baths[a]

Solvent	Bath Temperature (°C)
Carbon tetrachloride	−23°
Chlorobenzene	−45°
Chloroform	−63°
Ethyl acetate	−84°
Toluene	−95°
Methanol	−98°
Carbon Disulfide	−112°
Ethyl alcohol	−116°
n-Pentane	−131°
Isopentane	−160°

[a]Liquid N_2 is slowly poured into a Dewar flask containing the organic solvent. The mixture is stirred continuously during the addition to avoid forming a crust on the surface, that could break the Dewar. R. E. Rondeau, *J. Chem. Eng. Data, 11*, 124 (1966).

II. Heating Baths[a]

Bath Material	M.P. (°C)	B.P. (°C)
Water	0°	100°
Mineral oil[b]	—	—
Silicone oils[c]	—	—
Dibutyl phthalate	—	340°
40% $NaNO_2$, 7% $NaNO_3$, 53% KNO_3[d]	142°	—
51.3% KNO_3, 48.7% $NaNO_3$[d]	219°	—

[a]Normally these will be heated by an external source such as a hot plate. R. S. Egly, "Heating and Cooling," in Vol. 3, Part 2 of *Technique in Organic Chemistry*, A. Weissberger, Ed., Interscience, New York, 1957, p. 152.

[b]Useful up to approximately 180°.

[c]Very stable, but expensive. Available in several fractions with useful ranges from 30° to 280° from Dow Corning.

[d]Useful up to 500°.

APPENDIX 10   SI Units

The International Bureau of Weights and Measures has recommended an international system *(Systéme International)* for expressing quantitative units in a consistent and logical manner. It consists of several basic units (length, mass, time, etc.) and many derived units (volume, velocity, force, energy, etc.). Also, prefixes for the units are to be changed in steps of 10^3, which means that the use of units such as centimeter (10^{-2} meter) is not encouraged. Below are listed some basic units, derived units, and prefixes in the *Systéme International*. For a more complete description, see the references.

I. Basic SI Units

Physical Quantity	Name of Unit	Symbol
Length	meter	m
Mass	kilogram	kg
Time	second	s
Electric current	ampere	A
Thermodynamic temperature	degree Kelvin	K
Amount of substance	mole	mol

II. Some Derived SI Units

Physical Quantity	SI Name	SI Symbol
Area	square meter	m^2
Volume	cubic meter	m^3
Density	kilogram per cubic meter	$kg \cdot m^{-3}$
Velocity	meter per second	$m \cdot s^{-1}$
Acceleration	meter per second squared	$m \cdot s^{-2}$
Force	newton (N)	$kg \cdot m \cdot s^{-2} = J \cdot m^{-1}$
Pressure	newton per square meter	$N \cdot m^{-2}$
Energy	joule (J)	$kg \cdot m^2 \cdot s^{-2} = N \cdot m$
Power	watt (W)	$kg \cdot m^2 \cdot s^{-3} = J \cdot s^{-1}$
Electric charge	coulomb (C)	$A \cdot s$
Electric potential difference	volt (V)	$kg \cdot m^2 \cdot s^{-3} \cdot A^{-1} = J \cdot A^{-1} \cdot s^{-1}$
Electric resistance	ohm (Ω)	$kg \cdot m^2 \cdot s^{-3} \cdot A^{-2} = V \cdot A^{-1}$

III. Prefixes

	Prefix	Symbol
10^{12}	tera	T
10^9	giga	G
10^6	mega	M
10^3	kilo	k
10^{-3}	milli	m
10^{-6}	micro	μ
10^{-9}	nano	n
10^{-12}	pico	p

IV. SI Equivalents of Some Common Non-SI Units

Physical Quantity	Name	SI Equivalent
Length	angstrom (Å)	10^{-10} m
	micron (μ)	10^{-6} m
Volume	liter	10^{-3} m^3
Force	dyne (dyn)	10^{-5} N
Pressure	atmosphere (atm)	101,325 N·m^{-2}
	torr (mm Hg)	133.322 N·m^{-2}
Energy	calorie (cal)	4.1840 J
	electron volt (eV)	0.16021×10^{-18} J

REFERENCES

R. B. Heslop and G. M. Wild, *SI Units in Chemistry,* Applied Science Publishers Ltd., London, 1971.

A. C. Norris, "SI Units in Physico-chemical Calculations," *J. Chem. Educ., 48,* 797 (1971).

M. A. Paul, "The International System of Units—Development and Progress," *Journal of Chemical Documentation, 11,* 3 (1971).

"Policy for NBS Usage of SI Units," *J. Chem. Educ., 48,* 569 (1971).

T. I. Quickenden and R. C. Marshall, "Magnetochemistry in SI Units," *J. Chem. Educ., 49,* 114 (1972). Also see J. I. Hoppeé, *J. Chem. Educ., 49,* 505 (1972).

Notes to Instructor

To minimize the amount of time spent organizing equipment for the experiments, we have placed the equipment for each experiment in a certain padlocked drawer in the laboratory. The instructor opens the drawers at the beginning of the class, and locks them at the end of the period, while ensuring that the apparatus is clean and unbroken. This arrangement has saved a lot of time normally spent searching for apparatus. Since several students will use the same equipment throughout the duration of the course, it is absolutely essential that glassware be left clean and in good condition at the end of each laboratory period. As we have operated the laboratory, each student has his own drawer in which he keeps a minimum of equipment, such as spatulas, a wash bottle, suction-filtration apparatus, and so forth.

The following sources of supplies and equipment are simply those that we have found satisfactory. That we have included specific brand names in no way implies that these items are superior to others on the market.

Introduction

On our utility vacuum line, we have used 24/40 standard taper joints everywhere except on the cold trap, where a 45/50 joint is used. High-vacuum stopcocks (Figure 19–3) are used throughout the line. See Experiment 19.

Experiment 1

We have used an Industrial Instruments conductivity bridge, model RC 16B2.

The Harshaw Chemical Co., 6809 Cochran Road, Solon, Ohio 44139 produces relatively inexpensive NaCl plates in a variety of sizes suitable for almost any mull holder. The unpolished plates are satisfactory for mulls.

Experiment 2

Mineral oil or water covered with a layer of mineral oil has been used for the 60°C thermostatted bath. The temperature of the bath should be constant within ±0.2°C.

Experiment 3

Ground glass joints are not necessary in setting up the glass apparatus. Rubber stopper connections have performed well.

A thermocouple associated with a Thermolyne control unit was used to measure and control the temperature of our simple inexpensive tube furnace.

Experiment 4

Anhydrous $CrCl_3$ may be purchased commercially if it is not prepared in Experiment 3.

Use an ammonia cylinder that can be handled without difficulty. We have used a number 2 cylinder, which contains 15 lb of NH_3. Consult the manufacturer's catalog for the presence of an eductor tube in a given cylinder. While lecture bottles are very convenient for handling, they contain only 3/8 lb of NH_3, which will be consumed rapidly by a relatively large class.

Experiment 5

The Gouy apparatus was constructed from the following components:

The balance was a standard double-pan analytical balance having 0.1 mg sensitivity. A hole was drilled through the floor of the balance case to permit attachment of the silver chain to the balance pan. The silver chain was purchased at a jewelry store.

The Alpha Scientific Laboratories Model AL7500 electromagnet in conjunction with the AL7500 power supply was used without an auxiliary current regulator. Although a current regulator would obviate the need to continually adjust the current, the results were nevertheless very satisfactory. The magnet had pole caps of 4 inch diameter, which were separated by 3/4 of an inch. All measurements were made at the same field strength, which was about 7000 gauss.

Protecting the sample tube in the field from drafts with a glass tube as shown in Figure 5–1 was difficult because of the small pole gap. Instead, a plastic box was built that encased the whole magnet. The removal of the glass protection tube allowed for more convenient sample suspension.

The sample tubes were of the type shown in Figure 5-1. They were constructed from Pyrex tubing and had flat bottoms.

The calibration standards, $HgCo(NCS)_4$ and $[Ni(en)_3]S_2O_3$, and their syntheses are given in the following references:

B. N. Figgis and R. S. Nyholm, *J. Chem. Soc.,* 4190 (1958).

N. F. Curtis, *J. Chem. Soc.,* 3147 (1961).

Experiment 6

Anhydrous $CrCl_3$ is commercially available.

Experiment 7

The discarded Dowex 50W-X8 resin may be regenerated by passing 6N HCl through it. This is followed with a water wash until the effluent is no longer acidic.

Experiment 8

We have used a Rudolph Model 63 visual polarimeter with a sodium light source. The polarimeter tube had a 1 decimeter light path length and was filled through a central tube.

Experiment 9

We have used model X-17-17 glove bags purchased from Instruments for Research and Industry (see Experiment 11, Notes). PCl_5 and $SbCl_5$ are available from most large chemical suppliers. $P(C_6H_5)_3$ may be purchased from most organic chemical suppliers, including Aldrich Chemical Company (see Experiment 15, Notes).

Experiment 10

The amount of N_2O_4 prepared in this experiment is sufficient for the preparation of anhydrous $Cu(NO_3)_2$ as directed in Experiment 11. Infrared gas cells are available from infrared equipment suppliers.

Experiment 11

A lecture bottle cylinder of N_2O_4 is conveniently used in this preparation. Glove bags (dry bags) may be purchased from Instruments for Research and Industry, 108 Franklin Ave., Cheltenham, Pa. 19012. See Experiment 17 for supplier of silicone fluids for high temperature baths. Linde molecular sieves type 4A and Apiezon Q are available from Fisher Scientific Co., 711 Forbes Ave., Pittsburgh, Pa. 15219.

Experiment 13

We have used the relatively inexpensive Corning Model 5 pH meter with standard electrodes in this experiment with satisfactory results. The standardizing buffers were purchased from chemical suppliers. A calculator is recommended for evaluating the stability constants.

Experiment 14

$Mo(CO)_6$ and $Cr(CO)_6$ are available from Pressure Chemical Co., 3419–25 Smallman St., Pittsburgh, Pa. 15201, Strem Chemicals, Inc., 150 Andover St., Danvers, Mass., or Alfa Inorganics, Inc., 8 Congress St., Beverly, Mass. 01915.

We have routinely used the Parr bomb shown in Figure 14–2. It consists of a model 4753 general purpose ($1\frac{1}{2}$ inch × 7 inch deep, 200 ml capacity) type 316 stainless steel bomb, a model A427HC gauge block assembly with silver rupture disc, and model A495HC hose assembly. It is available from Parr Instrument Company, 211 53rd St., Moline, Illinois 61265. Reactions in the bomb were carried out in a glass liner (a test tube), which fit snugly into the bomb leaving about 2 cm between the top of the liner and the bomb head.

See Experiment 17 for suppliers of silicone fluids for high temperature baths.

Experiment 15

The sensitivity of $C_5H_5Fe(CO)_2CH_3$ to air makes its synthesis more challenging than other syntheses in this book.

Dicyclopentadiene purchased from Aldrich Chemical Co., 940 W. St. Paul Ave., Milwaukee, Wis. 53233, has worked well in this experiment. Samples obtained from other suppliers have not always behaved so well.

$Fe(CO)_5$ and $[C_5H_5Fe(CO)_2]_2$ may be obtained from suppliers listed above under Experiment 14.

Experiment 16

Ferrocene may be purchased from most large chemical supply houses. For TLC, we have used "Silica gel H for TLC" manufactured by E. Merck AG-Darmstadt (Germany) and distributed by Brinkmann Instruments Inc., Cantiague Rd., Westbury, New York 11590. For column chromatography, powdered silica gel "suitable for chromatographic use" from J. T. Baker Chemical Company, 222 Phillipsburg, N. J, was used.

Experiment 17

Anhydrous SnCl$_4$ is available from most inorganic chemical suppliers. Silicone fluids for high temperature baths are available from Dow Corning Corporation, Midland, Michigan 48640, and many supply houses.

Experiment 18

We have used an Electro Model D–612T filtered power supply (Electro Products Laboratories, Chicago, Ill.). See A. K. Vijh and R. S. Alwitt, *J. Chem. Educ.*, *46*, 121 (1969) for a convenient method of sealing metal wires into glass tubing.

Experiment 19

We have used a Welch Model 1405 vacuum pump in this experiment. Kel-F vacuum grease is available from most chemical equipment suppliers.

The vacuum line should be as sturdy as possible. Even then, breakage will occur; it is convenient to have a spare section of the line available. Virtually all breakage occurred when clamps were tightened too firmly or when keys in stopcocks were turned without proper support. Although the original vacuum line design has been retained in this second edition, we have modified our own line by using ball and socket rather than standard taper joints wherever possible. This has reduced breakage of the line considerably.

Although safety has been stressed in operating the vacuum line, we have had no injuries or even serious circumstances among the many students who have carried out this experiment. The laboratory has been briefly evacuated on occasion when BF$_3$ was liberated from a broken glass section. Also because of excessive BF$_3$ pressure in the line, BF$_3$ has occasionally forced the mercury out of the manometer, again freeing the gas into the laboratory. On the very few occasions when pressure has built up in Trap 1 with the inlet stopcock and S1 closed, one of these stopcocks has popped out, eliminating any danger of explosion.

Experiment 20

t-Butylamine was purchased from the Aldrich Chemical Co. (See Experiment 15, Notes). The $NaBH_4$ was purchased from Alfa Inorganics, Inc. (See Experiment 14, Notes.)

Index of Compounds ━━━━━━━━━

General Index ━━━━━━━━

Page numbers followed by (t) indicate tables.

235